Leveled Texts
for Mathematics

Data Analysis and Probability

Author

Stephanie Paris

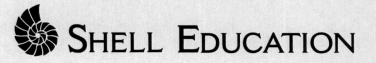

SHELL EDUCATION

Consultant

Barbara Talley, M.S.
Texas A&M University

Publishing Credits

Dona Herweck Rice, *Editor-in-Chief*; Lee Aucoin, *Creative Director*; Don Tran, *Print Production Manager;*
Sara Johnson, M.S.Ed., *Senior Editor*; Hillary Wolfe, *Editor;* Christina Hill, *Editor;*
Evelyn Garcia, *Associate Education Editor*; Neri Garcia, *Cover Designer;* Juan Chavolla, *Production Artist;*
Stephanie Reid, *Photo Editor;* Corinne Burton, M.S.Ed., *Publisher*

All images from Shutterstock.com

Shell Education

5301 Oceanus Drive

Huntington Beach, CA 92649

http://www.shelleducation.com

ISBN 978-1-4258-0755-9

©*2011 Shell Educational Publishing, Inc.*

Table of Contents

What Is Differentiation?

Over the past few years, classrooms have evolved into diverse pools of learners. Gifted students, English language learners, special needs students, high achievers, underachievers, and average students all come together to learn from one teacher. The teacher is expected to meet their diverse needs in one classroom. It brings back memories of the one-room schoolhouse during early American history. Not too long ago, lessons were designed to be one size fits all. It was thought that students in the same grade level learned in similar ways. Today, we know that viewpoint to be faulty. Students have differing learning styles, come from different cultures, experience a variety of emotions, and have varied interests. For each subject, they also differ in academic readiness. At times, the challenges teachers face can be overwhelming, as they struggle to figure out how to create learning environments that address the differences they find in their students.

What is differentiation? Carol Ann Tomlinson at the University of Virginia says, "Differentiation is simply a teacher attending to the learning needs of a particular student or small group of students, rather than teaching a class as though all individuals in it were basically alike" (2000). Differentiation can be carried out by any teacher who keeps the learners at the forefront of his or her instruction. The effective teacher asks, "What am I going to do to shape instruction to meet the needs of all my learners?" One method or methodology will not reach all students.

Differentiation encompasses what is taught, how it is taught, and the products students create to show what they have learned. When differentiating curriculum, teachers become the organizers of learning opportunities within the classroom environment. These categories are often referred to as content, process, and product.

- **Content:** Differentiating the content means to put more depth into the curriculum through organizing the curriculum concepts and structure of knowledge.

- **Process:** Differentiating the process means using varied instructional techniques and materials to enhance students' learning.

- **Product:** When products are differentiated, cognitive development and the students' abilities to express themselves improve, as they are given different product options.

Teachers should differentiate content, process, and product according to students' characteristics, including students' readiness, learning styles, and interests.

- **Readiness:** If a learning experience aligns closely with students' previous skills and understanding of a topic, they will learn better.

- **Learning styles:** Teachers should create assignments that allow students to complete work according to their personal preferences and styles.

- **Interests:** If a topic sparks excitement in the learners, then students will become involved in learning and better remember what is taught.

4

How to Differentiate Using This Product

According to the Common Core State Standards (2010), all students need to learn to read and discuss concepts across the content areas in order to be prepared for college and beyond. The leveled texts in this series help teachers differentiate mathematics content for their students to allow all students access to the concepts being explored. Each book has 15 topics, and each topic has a text written at four different reading levels. (See page 17 for more information.) While these texts are written at a variety of reading levels, all the levels remain strong in presenting the mathematics content and vocabulary. Teachers can focus on the same content standard or objective for the whole class, but individual students can access the content at their instructional reading levels rather than at their frustration levels.

Determining your students' instructional reading levels is the first step in the process. It is important to assess their reading abilities often so they do not get tracked into one level. Below are suggested ways to determine students' reading levels.

- **Running records:** While your class is doing independent work, pull your below-grade-level students aside, one at a time. Have them read aloud the lowest level of a text (the star level) individually as you record any errors they make on your own copy of the text. If students read accurately and fluently and comprehend the material, move them up to the next level and repeat the process. Following the reading, ask comprehension questions to assess their understanding of the material. Use your judgment to determine whether students seem frustrated as they read. As a general guideline, students reading below 90% accuracy are likely to feel frustrated as they read. There are also a variety of published reading assessment tools that can be used to assess students' reading levels with the running record format.

- **Refer to other resources:** Other ways to determine instructional reading levels include checking your students' Individualized Education Plans (IEPs), asking the school's resource teachers, or reviewing test scores. All of these resources should be able to give you the additional information you need to determine the reading level to begin with for your students.

Teachers can also use the texts in this series to scaffold the content for their students. At the beginning of the year, students at the lowest reading levels may need focused teacher guidance. As the year progresses, teachers can begin giving students multiple levels of the same text to allow them to work independently to improve their comprehension. This means each student would have a copy of the text at his or her independent reading level and instructional reading level. As students read the instructional-level texts, they can use the lower texts to better understand the difficult vocabulary. By scaffolding the content in this way, teachers can support students as they move up through the reading levels. This will encourage students to work with texts that are closer to the grade level at which they will be tested.

General Information About the Student Populations

Below-Grade-Level Students

By Dennis Benjamin

Gone are the days of a separate special education curriculum. Federal government regulations require that special needs students have access to the general education curriculum. For the vast majority of special needs students today, their Individualized Education Plans (IEPs) contain current and targeted performance levels but few short-term content objectives. In other words, the special needs students are required to learn the same content as their on-grade-level peers.

Be well aware of the accommodations and modifications written in students' IEPs. Use them in your teaching and assessment so they become routine. If you hold high expectations of success for all of your students, their efforts and performances will rise as well. Remember the root word of *disability* is *ability*. Go to the root needs of the learner and apply good teaching. The results will astound and please both of you.

English Language Learners

By Marcela von Vacano

Many school districts have chosen the inclusion model to integrate English language learners into mainstream classrooms. This model has its benefits as well as its drawbacks. One benefit is that English language learners may be able to learn from their peers by hearing and using English more frequently. One drawback is that these second-language learners cannot understand academic language and concepts without special instruction. They need sheltered instruction to take the first steps toward mastering English. In an inclusion classroom, the teacher may not have the time or necessary training to provide specialized instruction for these learners.

Acquiring a second language is a lengthy process that integrates listening, speaking, reading, and writing. Students who are newcomers to the English language are not able to process information until they have mastered a certain number of structures and vocabulary words. Students may learn social language in one or two years. However, academic language takes up to eight years for most students.

Teaching academic language requires good planning and effective implementation. Pacing, or the rate at which information is presented, is another important component in this process. English language learners need to hear the same word in context several times, and they need to practice structures to internalize the words. Reviewing and summarizing what was taught are absolutely necessary for English language learners.

General Information About the Student Populations

On-Grade-Level Students

By Wendy Conklin

On-grade-level students often get overlooked when planning curriculum. More emphasis is usually placed on those who struggle and, at times, on those who excel. Teachers spend time teaching basic skills and even go below grade level to ensure that all students are up to speed. While this is a noble endeavor and is necessary at times, in the midst of it all, the on-grade-level students can get lost in the shuffle. We must not forget that differentiated strategies are good for the on-grade-level students, too. Providing activities that are too challenging can frustrate these students; on the other hand, assignments that are too easy can be boring and a waste of their time. The key to reaching this population successfully is to find just the right level of activities and questions while keeping a keen eye on their diverse learning styles.

Above-Grade-Level Students

By Wendy Conklin

In recent years, many state and school district budgets have cut funding that has in the past provided resources for their gifted and talented programs. The push and focus of schools nationwide is proficiency. It is important that students have the basic skills to read fluently, solve problems, and grasp mathematical concepts. As a result, funding has been redistributed in hopes of improving test scores on state and national standardized tests. In many cases, the attention has focused only on improving low test scores to the detriment of the gifted students who need to be challenged.

Differentiating the products you require from your students is a very effective and fairly easy way to meet the needs of gifted students. Actually, this simple change to your assignments will benefit all levels of students in your classroom. While some students are strong verbally, others express themselves better through nonlinguistic representation. After reading the texts in this book, students can express their comprehension through different means, such as drawings, plays, songs, skits, or videos. It is important to identify and address different learning styles. By giving more open-ended assignments, you allow for more creativity and diversity in your classroom. These differentiated products can easily be aligned with content standards. To assess these standards, use differentiated rubrics.

Strategies for Using the Leveled Texts

Below-Grade-Level Students

By Dennis Benjamin

Vocabulary Scavenger Hunt

A valuable prereading strategy is a Vocabulary Scavenger Hunt. Students preview the text and highlight unknown words. Students then write the words on specially divided pages. The pages are divided into quarters with the following headings: *Definition*, *Sentence*, *Examples*, and *Nonexamples*. A section called *Picture* can be placed over the middle of the chart to give students a visual reminder of the word and its definition.

Example Vocabulary Scavenger Hunt

estimate

Definition	Sentence
to determine roughly the size or quantity	We need to estimate how much paint we will need to buy.
Examples	**Nonexamples**
estimating a distance; estimating an amount	measuring to an exact number; weighing and recording an exact weight

This encounter with new vocabulary enables students to use it properly. The definition identifies the word's meaning in student-friendly language. The sentence should be written so that the word is used in context. This helps the student make connections with background knowledge. Illustrating the sentence gives a visual clue. Examples help students prepare for factual questions from the teacher or on standardized assessments. Nonexamples help students prepare for *not* and *except for* test questions such as "All of these are polygons *except for*…" and "Which of these terms in this expression is not a constant?" Any information the student was unable to record before reading can be added after reading the text.

Strategies for Using the Leveled Texts *(cont.)*

Below-Grade-Level Students *(cont.)*

Graphic Organizers to Find Similarities and Differences

Setting a purpose for reading content focuses the learner. One purpose for reading can be to identify similarities and differences. This is a skill that must be directly taught, modeled, and applied. The authors of *Classroom Instruction That Works* state that identifying similarities and differences "might be considered the core of all learning" (Marzano, Pickering, and Pollock 2001). Higher-level tasks include comparing and classifying information and using metaphors and analogies. One way to scaffold these skills is through the use of graphic organizers, which help students focus on the essential information and organize their thoughts.

Example Classifying Graphic Organizer

Equation	Constants	Variables	Number of Terms
$7 + 12 = 19$	7, 12, 19	none	3
$3x = 12$	12	x	2
$a + b$	none	a, b	2
$a^2 + b^2 + c^2$	none	a, b, c	3

The Riddles Graphic Organizer allows students to compare and contrast two-dimensional shapes using riddles. Students first complete a chart you've designed. Then, using that chart, they can write summary sentences. They do this by using the riddle clues and reading across the chart. Students can also read down the chart and write summary sentences. With the chart below, students could write the following sentences: A circle is not a polygon. The interior angles of a triangle always add up to 180°.

Example Riddles Graphic Organizer

What Am I?	Circle	Square	Triangle	Rectangle
I come in different configurations.			x	x
I am a polygon.		x	x	x
I am a closed shape.	x	x	x	x
My interior angles always add up to 180°.			x	
I have at least three vertices.		x	x	x

Strategies for Using the Leveled Texts *(cont.)*

Below-Grade-Level Students *(cont.)*

Framed Outline

This is an underused technique that bears great results. Many below-grade-level students have problems with reading comprehension. They need a framework to help them attack the text and gain confidence in comprehending the material. Once students gain confidence and learn how to locate factual information, the teacher can fade out this technique.

There are two steps to successfully using this technique. First, the teacher writes cloze sentences. Second, the students complete the cloze activity and write summary sentences.

Example Framed Outline

A _____ graph is used to show how two variables may be related to each other. A graph should have a _____. The axes should be labeled and a proper scale should be shown.

Summary Sentences

A good graph should correctly show the data. It should have a title; the axes should be labeled correctly; it should show the proper scale.

Modeling Written Responses

A frequent criticism heard by special educators is that below-grade-level students write poor responses to content-area questions. This problem can be remedied if resource and classroom teachers model what good answers look like. While this may seem like common sense, few teachers take the time to do this. They just assume all children know how to respond in writing.

This is a technique you may want to use before asking your students to respond to the You Try It questions associated with the leveled texts in this series. First, read the question aloud. Then, write the question on the board or an overhead and think aloud about how you would go about answering the question. Next, solve the problem showing all the steps. Introduce the other problems and repeat the procedure. Have students explain how they solved the problems in writing so that they make the connection that quality written responses are part of your expectations.

Strategies for Using the Leveled Texts (cont.)

English Language Learners

By Marcela von Vacano

Effective teaching for English language learners requires effective planning. In order to achieve success, teachers need to understand and use a conceptual framework to help them plan lessons and units. There are six major components to any framework. Each is described in detail below.

1. Select and Define Concepts and Language Objectives—Before having students read one of the texts in this book, the teacher must first choose a mathematical concept and language objective (reading, writing, listening, or speaking) appropriate for the grade level. Then, the next step is to clearly define the concept to be taught. This requires knowledge of the subject matter, alignment with local and state objectives, and careful formulation of a statement that defines the concept. This concept represents the overarching idea. The mathematical concept should be written on a piece of paper and posted in a visible place in the classroom.

By the definition of the concept, post a set of key language objectives. Based on the content and language objectives, select essential vocabulary from the text. The number of new words selected should be based on students' English language levels. Post these words on a word wall that may be arranged alphabetically or by themes.

2. Build Background Knowledge—Some English language learners may have a lot of knowledge in their native language, while others may have little or no knowledge. The teacher will want to build the background knowledge of the students using different strategies such as the following:

> **Visuals:** Use posters, photographs, postcards, newspapers, magazines, drawings, and video clips of the topic you are presenting.

> **Realia:** Bring real-life objects to the classroom. If you are teaching about measurement, bring in items such as thermometers, scales, time pieces, and rulers.

> **Vocabulary and Word Wall:** Introduce key vocabulary in context. Create families of words. Have students draw pictures that illustrate the words and write sentences about the words. Also, be sure you have posted the words on a word wall in your classroom.

> **Desk Dictionaries:** Have students create their own desk dictionaries using index cards. On one side, they should draw a picture of the word. On the opposite side, they should write the word in their own language and in English.

Strategies for Using the Leveled Texts *(cont.)*

English Language Learners *(cont.)*

3. Teach Concepts and Language Objectives—The teacher must present content and language objectives clearly. He or she must engage students using a hook and must pace the delivery of instruction, taking into consideration students' English language levels. The concept or concepts to be taught must be stated clearly. Use the first languages of the students whenever possible or assign other students who speak the same languages to mentor and to work cooperatively with the English language learners.

Lev Semenovich Vygotsky, a Russian psychologist, wrote about the Zone of Proximal Development (ZPD). This theory states that good instruction must fill the gap that exists between the present knowledge of a child and the child's potential (1978). Scaffolding instruction is an important component when planning and teaching lessons. English language learners cannot jump stages of language and content development. You must determine where the students are in the learning process and teach to the next level using several small steps to get to the desired outcome. With the leveled texts in this series and periodic assessment of students' language levels, teachers can support students as they climb the academic ladder.

4. Practice Concepts and Language Objectives—English language learners need to practice what they learn with engaging activities. Most people retain knowledge best after applying what they learn to their own lives. This is definitely true for English language learners. Students can apply content and language knowledge by creating projects, stories, skits, poems, or artifacts that show what they learned. Some activities should be geared to the right side of the brain. For students who are left-brain dominant, activities such as defining words and concepts, using graphic organizers, and explaining procedures should be developed. The following teaching strategies are effective in helping students practice both language and content:

> **Simulations:** Students learn by doing. For example, when teaching about data analysis, have students do a survey about their classmates' favorite sports. First, students make a list of questions and collect the necessary data. Then, they tally the responses and determine the best way to represent the data. Lastly, students create a graph that shows their results and display it in the classroom.

> **Literature response:** Read a text from this book. Have students choose two concepts described or introduced in the text. Ask students to create a conversation two people might have to debate which concept is useful. Or, have students write journal entries about real-life ways they use these mathematical concepts.

#50755—Leveled Texts for Mathematics: Data Analysis and Probability   © Shell Education

English Language Learners *(cont.)*

4. Practice Concepts and Language Objectives *(cont.)*

Have a short debate: Make a controversial statement such as "It isn't necessary to learn addition." After reading a text in this book, have students think about the question and take a position. As students present their ideas, one student can act as a moderator.

Interview: Students may interview a member of the family or a neighbor in order to obtain information regarding a topic from the texts in this book. For example: What are some ways you use geometry in your work?

5. Evaluation and Alternative Assessments—We know that evaluation is used to inform instruction. Students must have the opportunity to show their understanding of concepts in different ways and not only through standard assessments. Use both formative and summative assessments to ensure that you are effectively meeting your content and language objectives. Formative assessment is used to plan effective lessons for a particular group of students. Summative assessment is used to find out how much the students have learned. Other authentic assessments that show day-to-day progress are: text retelling, teacher rating scales, student self-evaluations, cloze testing, holistic scoring of writing samples, performance assessments, and portfolios. Periodically assessing student learning will help you ensure that students continue to receive the correct levels of texts.

6. Home-School Connection—The home-school connection is an important component in the learning process for English language learners. Parents are the first teachers, and they establish expectations for their children. These expectations help shape the behavior of their children. By asking parents to be active participants in the education of their children, students get a double dose of support and encouragement. As a result, families become partners in the education of their children and chances for success in your classroom increase.

You can send home copies of the texts in this series for parents to read with their children. You can even send multiple levels to meet the needs of your second-language parents as well as your students. In this way, you are sharing your mathematics content standards with your whole second-language community.

Strategies for Using
the Leveled Texts *(cont.)*

Above-Grade-Level Students

By Wendy Conklin

Open-Ended Questions and Activities

Teachers need to be aware of activities that provide a ceiling that is too low for gifted students. When given activities like this, gifted students become bored. We know these students can do more, but how much more? Offering open-ended questions and activities will give high-ability students the opportunities to perform at or above their ability levels. For example, ask students to evaluate mathematical topics described in the texts, with questions such as "Do you think students should be allowed to use calculators in math?" or "What do you think you would need to build a two-story dog house?" These questions require students to form opinions, think deeply about the issues, and form several different responses in their minds. To questions like these, there really is no single correct answer.

The generic, open-ended question stems listed below can be adapted to any topic. There is one You Try It question for each topic in this book. Use questions or statements like the ones shown here to develop further discussion for the leveled texts.

- In what ways did…
- How might you have done this differently…
- What if…
- What are some possible explanations for…
- How does this affect…
- Explain several reasons why…
- What problems does this create…
- Describe the ways…
- What is the best…
- What is the worst…
- What is the likelihood…
- Predict the outcome…
- Form a hypothesis…
- What are three ways to classify…
- Support your reason…
- Make a plan for…
- Propose a solution…
- What is an alternative to…

Strategies for Using the Leveled Texts *(cont.)*

Above-Grade-Level Students *(cont.)*

Student-Directed Learning

Because they are academically advanced, above-grade-level students are often the leaders in classrooms. They are more self-sufficient learners, too. As a result, there are some student-directed strategies that teachers can employ successfully with these students. Remember to use the texts in this book as jump starts so that students will be interested in finding out more about the mathematical concepts presented. Above-grade-level students may enjoy any of the following activities:

- Writing their own questions, exchanging their questions with others, and grading the responses.
- Reviewing the lesson and teaching the topic to another group of students.
- Reading other nonfiction texts about these mathematical concepts to further expand their knowledge.
- Writing the quizzes and tests to go along with the texts.
- Creating illustrated timelines to be displayed as visuals for the entire class.
- Putting together multimedia presentations about the mathematical concepts.

Tiered Assignments

Teachers can differentiate lessons by using tiered assignments, or scaffolded lessons. Tiered assignments are parallel tasks designed to have varied levels of depth, complexity, and abstractness. All students work toward one goal, concept, or outcome, but the lesson is tiered to allow for different levels of readiness and performance. As students work, they build on their prior knowledge and understanding. Students are motivated to be successful according to their own readiness and learning preferences.

Guidelines for writing tiered lessons include the following:

1. Pick the skill, concept, or generalization that needs to be learned.
2. Think of an on-grade-level activity that teaches this skill, concept, or generalization.
3. Assess the students using classroom discussions, quizzes, tests, or journal entries and place them in groups.
4. Take another look at the activity from Step 2. Modify this activity to meet the needs of the below-grade-level and above-grade-level learners in the class. Add complexity and depth for the above-grade-level students. Add vocabulary support and concrete examples for the below-grade-level students.

How to Use This Product

Readability Chart

Title of the Text	Star	Circle	Square	Triangle
Collecting Data	2.2	3.5	5.0	6.6
Creating Pictographs	2.2	3.3	5.3	6.5
Analyzing Pictographs	2.1	3.4	5.2	6.8
Creating Bar Graphs	1.9	3.4	5.0	6.8
Analyzing Bar Graphs	2.2	3.5	5.3	6.6
Creating Line Graphs	2.2	3.5	5.0	6.6
Analyzing Line Graphs	2.2	3.5	5.2	6.5
Creating Circle Graphs	2.2	3.4	5.3	6.5
Analyzing Circle Graphs	2.2	3.1	5.0	6.5
Comparing Graphs	1.9	3.0	5.0	6.5
What Does *Mean* Mean?	2.2	3.5	5.1	6.6
Median in the Middle	2.1	3.5	5.1	6.5
Mode and Range	2.2	3.0	5.1	6.5
Probability of Events	2.2	3.5	5.0	6.6
Probability Experiments	2.2	3.1	5.3	6.6

Components of the Product

Strong Image Support

- Each level of text includes multiple primary sources. These images, photographs, and illustrations add interest to the texts. The mathematical images also serve as visual support for second-language learners. They make the texts more context-rich and bring the examples to life.

How to Use This Product *(cont.)*

Components of the Product *(cont.)*

Practice Problems

- The introduction often includes a challenging question or riddle. The answer can be found on the next page at the end of the lesson.
- Each level of text includes a You Try It section where the students are asked to solve problems using the skill or concept discussed in the text.
- Although the mathematics is the same, the questions may be worded slightly differently depending on the reading level of the passage.

The Levels

- There are 15 topics in this book. Each topic is leveled to four different reading levels. The images and fonts used for each level within a topic look the same.
- Behind each page number, you'll see a shape. These shapes indicate the reading levels of each text so that you can make sure students are working with the correct texts. The reading levels fall into the ranges indicated below. See the chart on page 16 for the specific reading levels of each lesson.

Leveling Process

- The texts in this series were originally authored by mathematics educators. A reading expert went through the texts and leveled each one to create four distinct reading levels.
- A mathematics expert then reviewed each passage for accuracy and mathematical language.
- The texts were then leveled one final time to ensure the editorial changes made during the process kept them within the ranges described to the left.

Levels
1.5–2.2

Levels
3.0–3.5

Levels
5.0–5.5

Levels
6.5–7.2

How to Use This Product *(cont.)*

Tips for Managing the Product

How to Prepare the Texts

- When you copy these texts, be sure you set your copier to copy photographs. Run a few test pages and adjust the contrast as necessary. If you want the students to be able to appreciate the images, you need to carefully prepare the texts for them.

- You also have full-color versions of the texts provided in PDF form on the CD. (See page 142 for more information.) Depending on how many copies you need to make, printing the full-color versions and copying those might work best for you.

- Keep in mind that you should copy two-sided to two-sided if you pull the pages out of the book. The shapes behind the page numbers will help you keep the pages organized as you prepare them.

Distributing the Texts

Some teachers wonder about how to hand the texts out within one classroom. They worry that students will feel insulted if they do not get the same papers as their neighbors. The first step in dealing with these texts is to set up your classroom as a place where all students learn at their individual instructional levels. Making this clear as a fact of life in your classroom is key. Otherwise, the students may constantly ask about why their work is different. You do not need to get into the technicalities of the reading levels. Just state it as a fact that every student will not be working on the same assignment every day. If you do this, then passing out the varied levels is not a problem. Just pass them to the correct students as you circle the room.

If you would rather not have students openly aware of the differences in the texts, you can try these ways to pass out the materials:

- Make a pile in your hands from star to triangle. Put your finger between the circle and square levels. As you approach each student, you pull from the top (star), above your finger (circle), below your finger (square), or the bottom (triangle). If you do not hesitate too much in front of each desk, the students will probably not notice.

- Begin the class period with an opening activity. Put the texts in different places around the room. As students work quietly, circulate and direct students to the right locations for retrieving the texts you want them to use.

- Organize the texts in small piles by seating arrangement so that when you arrive at a group of desks you have just the levels you need.

How to Use This Product *(cont.)*

Correlation to Mathematics Standards

Shell Education is committed to producing educational materials that are research and standards based. In this effort, we have correlated all of our products to the academic standards of all 50 United States, the District of Columbia, the Department of Defense Dependent Schools, and all Canadian provinces. We have also correlated to the Common Core State Standards.

How to Find Standards Correlations

To print a customized correlation report of this product for your state, visit our website at **http://www.shelleducation.com** and follow the on-screen directions. If you require assistance in printing correlation reports, please contact Customer Service at 1-877-777-3450.

Purpose and Intent of Standards

Legislation mandates that all states adopt academic standards that identify the skills students will learn in kindergarten through grade twelve. Many states also have standards for Pre-K. This same legislation sets requirements to ensure the standards are detailed and comprehensive.

Standards are designed to focus instruction and guide adoption of curricula. Standards are statements that describe the criteria necessary for students to meet specific academic goals. They define the knowledge, skills, and content students should acquire at each level. Standards are also used to develop standardized tests to evaluate students' academic progress. Teachers are required to demonstrate how their lessons meet state standards. State standards are used in the development of all of our products, so educators can be assured they meet the academic requirements of each state.

TESOL Standards

The lessons in this book promote English language development for English language learners. The standards listed on the Teacher Resource CD support the language objectives presented throughout the lessons.

NCTM Standards Correlation Chart

The chart on the next page shows the correlation to the National Council for Teachers of Mathematics (NCTM) standards. This chart is also available on the Teacher Resource CD (*nctm.pdf*).

NCTM Standards

NCTM Standard	Lesson	Page
Design investigations to address a question and consider how data-collection methods affect the nature of the data set	Collecting Data	21–28
Collect data using observations, surveys, and experiments	Collecting Data	21–28
Represent data using tables and graphs such as line plots, bar graphs, and line graphs	Creating Pictographs; Creating Bar Graphs; Creating Line Graphs; Creating Circle Graphs	29–36, 45–52, 61–68, 77–84
Recognize the differences in representing categorical and numerical data	Comparing Graphs	93–100
Describe the shape and important features of a set of data and compare related data sets, with an emphasis on how the data are distributed	Analyzing Pictographs; Analyzing Bar Graphs; Analyzing Line Graphs; Analyzing Circle Graphs; Comparing Graphs	37–44, 53–60, 69–76, 85–100
Use measures of center, focusing on the median, and understand what each does and does not indicate about the data set	Median in the Middle	109–116
Compare different representations of the same data and evaluate how well each representation shows important aspects of the data	Comparing Graphs; What Does *Mean* Mean?; Median in the Middle; Mode and Range	93–124
Propose and justify conclusions and predictions that are based on data and design studies to further investigate the conclusions or predictions	Analyzing Pictographs; Analyzing Bar Graphs; Analyzing Line Graphs; Analyzing Circle Graphs	37–44, 53–60, 69–76, 85–92
Describe events as likely or unlikely and discuss the degree of likelihood using such words as *certain*, *equally likely*, and *impossible*	Probability of Events; Probability Experiments	125–140
Predict the probability of outcomes of simple experiments and test the predictions	Probability of Events; Probability Experiments	125–140
Understand that the measure of the likelihood of an event can be represented by a number from 0 to 1	Probability of Events; Probability Experiments	125–140

Standards are listed with the permission of the National Council of Teachers of Mathematics (NCTM). NCTM does not endorse the content or validity of these alignments.

Collecting Data

Jeremy and Michelle do not agree on sports. Jeremy thinks that more people in their class like baseball. Michelle thinks more people like football. They want to learn the truth. How can they find out?

Basic Facts

First, they need to find out what sport their classmates like best. This is called **collecting data**. *Data* means "facts" or "pieces of information." In this case, they want to get data about their classmates. They need to find out which sport each person likes the best.

How to Collect Data

There are a lot of ways to collect data. You need to choose which way you want to use. Let us talk about four of the most common ways.

Observation

To **observe** means to notice. A good example of this is when you keep score in a game. Maybe you want to know how many points each team scores at a volleyball match. The best way to get this data is to watch the match! You can record each point as the teams make it. At the end of the game you will have all the data you need.

Survey

You have probably seen **surveys** done on the news or television shows. A survey is a way to collect data. You ask many people the same questions. Then you record their answers.

Experimentation

Sometimes you will not have the data you need, or the data is not easy to find. You may need to do an **experiment**. You need to find the data you want. Let us say that you want to know which kind of ball bounces the highest. The best way to find this data might be to try it. You could bounce each ball a few times. Then you could record how high they go. At the end, you could compare the results.

© Shell Education   #50755—*Leveled Texts for Mathematics: Data Analysis and Probability*

How to Collect Data *(cont.)*

Research

Many times you will have the data that you need. But the data is not in the form that you need. You need to do **research**. Maybe you want to know if taller basketball teams score more points. You would need to look at the height of the players. Then you would need to look at their number of wins. There are charts that show how tall each player is. And there are charts that show which teams won each game. But, there may not be a chart that compares the two. You could use both charts to find your data. Then you could create your own chart and compare the two.

Collecting Data in Our Daily Lives

People have been recording data for a long time. Early humans drew on cave walls to record data. These drawings are very old. They are from many years ago. The Incas had a smart way of keeping track of data. They would tie knots into strings. These knots would show data. They were called *quipus*. They were not easy to make. And they were hard to read. So they had people called *quipucamayu*. They made and read the quipus.

These days many people have jobs finding data. They also record data. Think of a library. The books are full of data. Someone had to write down all that data. Now think of the Internet. There is a lot of data that you can find. Someone had to collect all of it! Have you ever taken a survey? Survey takers are people whose job is to ask questions. Then they write down the answers. Scientists have to collect data, too. Think about reporters. They also have to collect data. Then they find ways to share it!

You Try It

Take a survey of your class. Think of a question to ask. The question should have two or three answers. An example might be, "Which fruit do you like best? Apples, oranges, or bananas?" Write the answers in a chart. Ask each person in the class your question. Only ask each person once. Make tally marks to keep track of the answers. Make one mark for each answer.

22

Collecting Data

Jeremy and Michelle do not agree about sports. Jeremy says that baseball is the most popular sport in their class. Michelle is sure that the most popular sport is football. They want to learn the truth. How can they find out?

Basic Facts

The first thing that Jeremy and Michelle need to do is get more information. In mathematics, we call this **collecting data**. *Data* means "facts" or "pieces of information." In this case, they want to collect data about their classmates. They need to find out which sport each person likes best.

How to Collect Data

The first step to gathering data is deciding which method to use. There are a lot of ways to collect data. Let us talk about four of the most common.

Observation

To **observe** means to notice. A good example of this is when you keep score in a game. Let us say you want to know how many points each team scores during a volleyball match. The best way to get this data is to watch the match! You can record each point as the teams make it. At the end of the game, you will have all the data from that match.

Survey

You have probably seen **surveys** done on the news or television shows. A survey is a way to collect data. In surveys you ask many people the same questions. Then you record their answers.

Experimentation

Sometimes you will not have the data you need, or the data is not easy to find. You may need to do an **experiment**. You need to gather the data you want. Let us say that you want to know which kind of ball bounces the highest. The best way to gather this data might be to try it. You could bounce each ball a few times. Then you could record how high they go. At the end, you could compare the results.

(23)

How to Collect Data (cont.)

Research

Many times the data that you need does exist. But the data is not in the form that you need. You need to do **research**. For instance, maybe you want to know if taller basketball teams score more points. You would need to look at the average height of the players. Then you would need to look at their number of wins. There are records that show how tall each player is. There are records that show which teams won each game. But, there may not be a chart that compares the two. Gather what you need by using the data on these two charts. Then, you could create your own chart and compare.

Collecting Data in Our Daily Lives

People have been collecting and recording data for a long time. Early humans made drawings on cave walls to record data. Some drawings are known to be 20,000 years old! The ancient Incas had a clever method for recording data. They would tie knots into strings. These knots would show important information. These knot-based records were called *quipus*. They were not easy to make. And they were not easy to read. There were special people called *quipucamayu*. They made and read the quipus.

These days many people have jobs collecting and recording data. Think of a library. All the data in those books was collected by someone. Now think of the Internet. All of the data that you can search for was collected by someone! Have you ever taken a survey? Survey takers are people whose job is to ask questions. Then they record the answers. Most scientists spend their time collecting data and recording it, too. And if you think about it, reporters gather data, too. Then they find ways to share it!

You Try It

Take a survey of your class. Think of a question to ask. The question should have two or three possible answers. An example might be, "Which fruit do you like best? Apples, oranges, or bananas?" Write the answers in a chart. Ask each person in the class your question. Only ask each person once. Make tally marks to keep track of the answers. Make one mark for each answer.

Collecting Data

Jeremy and Michelle are having a disagreement. Jeremy says that baseball is the most popular sport in their class. Michelle is confident that the most popular sport is football. How can the two friends find the truth?

Basic Facts

The first thing that Jeremy and Michelle must do is to get more information. In mathematics, we call this **collecting data**. *Data* means "facts" or "pieces of information." In this case, Jeremy and Michelle want to collect data about their classmates. They need to find out which sport each person likes more.

How to Collect Data

The first step to gathering data is deciding which method to use. There are many ways to collect data. Let us talk about four of the most common.

Observation

To **observe** means to notice. A good example of this is when you keep score in a game. Let us say you want to know how many points each team scores during a volleyball match. The best way to get this data is to watch the match! You can record each point as the teams make it. At the end of the game you will have all the data from that match.

Survey

If you ever watch the news or television game shows you have probably heard about surveys. A **survey** is a way to collect data. In surveys you ask many people the same question or questions. Then you record their answers.

Experimentation

Sometimes the information that you need does not already exist. Or the data is not easy to find. In these cases, you may need to do an **experiment** to gather the data you want. Let us say that you want to know which kind of ball bounces the highest. The best way to gather this data might be to try it out. You could bounce each ball several times. Then you could record how high they go. At the end, you could compare the results.

25

How to Collect Data (cont.)

Research

Many times the data that you need does exist. But it is not in the form that you need. You need to do **research**. For instance, let us say you wanted to know if taller basketball teams score more points. You would need to compare the average height of the players on each team with their number of wins per season. There are records that show how tall each player is. And there are records that show which teams won each game. But, there may not be a chart that compares the two. You could gather the data that you need by researching the information that is already recorded. Then, you could create your own chart and make the comparison.

Collecting Data in Our Daily Lives

People have been collecting and recording data for just about as long as there have been people. Early humans made drawings on cave walls to record information. Some drawings are known to be around 20,000 years old! The ancient Incas had an ingenious method for recording data that they collected. They would tie knots into strings to show payments, purchases, and other important information. These knot-based records were called *quipus*. They were not easy to make. They were not even easy to read. Special people called *quipucamayu* were in charge of making and reading the quipus.

These days many people make their living collecting and recording data. Think of a library. All the information in those books was collected by someone. Now think of the Internet. All of the information that you can search for was collected by someone! Have you ever taken a survey? Survey takers are people whose entire job is to ask questions and record the answers. Most scientists spend their time collecting information and recording it, too. And if you think about it, reporters spend all their time gathering information and then finding ways to share it!

You Try It

Conduct a survey with your class. Think of a question with two or three possible answers. An example might be, "Which fruit do you prefer: apples, oranges, or bananas?" Write the answers in a chart. Ask your question of each person in the class. Ask each person only once. Make tally marks to keep track of people's answers. Make one mark for each answer.

Collecting Data

Jeremy and Michelle are having a disagreement. Jeremy says that baseball is the most popular sport in their class. Michelle is certain that it is football. How can the two friends find the truth?

Basic Facts

First, Jeremy and Michelle need to collect information. In mathematics, we call this **collecting data**. *Data* means "factual information." In this case, Jeremy and Michelle want to collect data about which sport their classmates prefer, baseball or football.

How to Collect Data

There are many methods of collecting data. The first step in collecting data is deciding which of these to use. There are four common methods of data collection.

Observation

To **observe** means to watch carefully with attention to detail. A good example of observation is when you keep score in a game. Let us say you want to find out how many points each team scores during a volleyball match. The best way to collect this data is to watch the match! You can record each point as the teams make it and at the end of the game you will have all the data you need from that match.

Survey

If you ever watch the news or television game shows then you probably know about surveys. In **surveys** you ask many people the same question or questions and then you record their answers. You can then use the data that you collected to learn more about the topic you are researching.

Experimentation

Sometimes the information that you need does not already exist or is not easy to locate. In these cases, you may need to do an **experiment** to collect the data you want. Maybe you want to conduct an experiment to determine which kind of ball bounces the highest. For this experiment, the best way to collect data might be to try it out. You could bounce each ball several times and then record how high each ball bounced. At the end of the experiment, you could compare the results.

27

How to Collect Data (cont.)

Research

Many times the data that you need does exist. However, it is not in the form that you need. You need to do **research**. For instance, let us say you wanted to know if taller basketball teams score more points. You would need to compare the average height of the players on each team with their number of wins per season. There are records that show how tall each player is, and there are records that show which teams won each game. But, there may not be a chart that compares the two. You could gather the data that you need by researching the information that is already recorded. Then, you could create your own chart and make the comparison.

Collecting Data in Our Daily Lives

People have been collecting and recording data for just about as long as there have been people. Early humans made drawings on cave walls to record information. Some drawings are known to be around 20,000 years old! The ancient Incas had an ingenious method for recording data that they collected. They would tie knots into strings to show payments, purchases, and other important information. These knot-based records were called *quipus*. These quipus were difficult to make and were not easy to read. Special people called *quipucamayu* were in charge of making and reading the quipus.

There are many professions today where people make a living collecting and recording data. All the information in the books in a library was collected by someone. Even all of the information that you can search for on the Internet was collected by someone! Have you ever taken a survey? Survey takers are people whose entire job is to ask questions and record the answers. Most scientists spend their time collecting information and recording it, too. And if you think about it, reporters spend all their time gathering information and then finding ways to share it!

You Try It

Conduct a survey with your class. Ask your classmates a question with two or three possible answers. For example, "Which fruit do you prefer: apples, oranges, or bananas?" Write your classmates' answers in a chart. Make sure you ask each person in the class only once and make one tally mark for each classmate's answer.

Creating Pictographs

Marvin and Zuna wanted to keep track of the weather. They did this for the month of October. Every morning they would check outside. Then they would draw a picture of the weather. The chart they made is below. Why do you think there are no pictures in the "snowing" row?

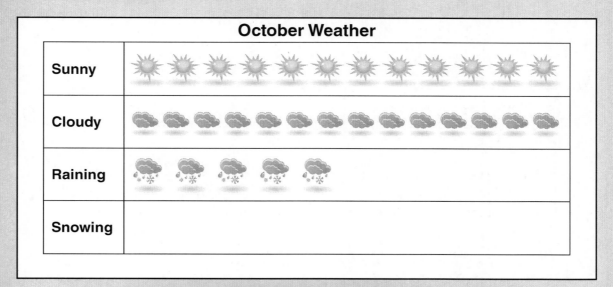

Basic Facts

One way to show data is to use a **pictograph**. Pictographs are picture graphs. They use pictures to show data. They are also called *pictograms*. Or, they are called *picture graphs*. Each picture in the chart can stand for one piece of data. Or, it can stand for a whole group. In the chart above, they used a picture to show each day. There are no pictures in the "snowing" row. It did not snow that month.

When a Picture Is Worth More Than a Piece of Data

Pictographs can be used to show a lot of data. It may not be a good idea to use one picture for each item. You might want each picture to show a group. Let us say that Mrs. Lundy's class wants to make a pictograph. They want to show what lunches people choose. The school has 300 people. So that would be a lot of drawings! Each picture stands for 20 meals instead of one meal. The key on the graph tells us this. That lets the chart stay much smaller. And it is easier to read.

But, what if the total number does not equal a group of 20? Pictographs can use parts of pictures. These pictures show smaller amounts. For example, half a picture would be 10 lunches.

But, what if the partial picture is meant to show 9 lunches? Or 11 lunches? It is hard to divide a picture to show this. We might not know the right amounts for sure. The person who made the graph must tell us. That is one of the drawbacks of a pictograph.

29

Creating a Pictograph

David wants to show how many families in his school have cats and dogs. He finds out that 208 families have cats. There are 160 families that have dogs. He wants to make a pictograph to show this data.

Step 1: Decide how many items each picture will represent. Often people choose factors of 10. But, different numbers make sense, too. A good choice might be to use a common factor. David notices that 208 and 160 are both divisible by 16. $208 \div 16 = 13$ and $160 \div 16 = 10$. He chooses to make each picture equal to 16 families. He notes his choice in a key.

Step 2: Decide on a picture. It should be something that looks like your data. You will need to use it many times. You might want to draw your pictures by hand. If so, choose something simple. David chooses to use a dog and cat! He labels the rows to show what they stand for.

Step 3: Create your graph. Draw pictures to show all the data. Make sure the partial pictures show the right amount. Be neat and clear. Make sure your pictures are the same size. If they are not, it will be hard to see the contrast.

Step 4: Give your graph a descriptive name and key. The title should tell you what the graph shows. The key should show you how much data each picture stands for. You can choose to leave out the key. But you can only do this if each picture shows one piece of data.

How Many Families Have Dogs and Cats as Pets	
Key: 1 picture = 16 families	
Dogs	🐕🐕🐕🐕🐕🐕🐕🐕🐕🐕
Cats	🐈🐈🐈🐈🐈🐈🐈🐈🐈🐈🐈🐈🐈

Pictographs in Our Daily Lives

Many newspapers use pictographs. And magazines use them, too. Pictographs are eye-catching. And they are easy to understand. They can be colorful. And they are a great way to show contrast. This makes them an ideal tool. They can help make data easy to understand.

You Try It

Create a pictograph. Show the favorite colors of the kids in your class.

Creating Pictographs

Marvin and Zuna wanted to keep track of the weather in their area. They did this for the month of October. Every morning they would check outside. Then they would draw a picture of what the weather was like. The chart they made is below. Why do you think there are no pictures in the "snowing" row?

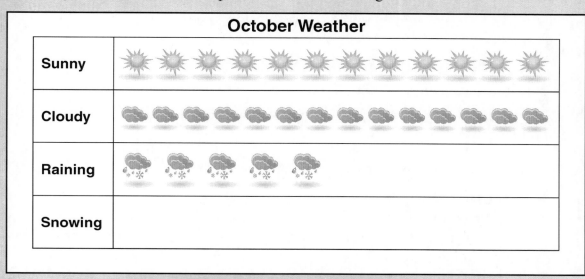

October Weather	
Sunny	☀☀☀☀☀☀☀☀☀☀☀☀
Cloudy	☁☁☁☁☁☁☁☁☁☁☁☁☁☁
Raining	🌧🌧🌧🌧🌧
Snowing	

Basic Facts

One way to show data is to use a **pictograph**. A pictograph is a graph that uses pictures to show the data. Sometimes they are called *pictograms* or *picture graphs*. Each picture in the chart can stand for one piece of data. Or, it can stand for a group of data points. In the example above, the children used a picture to show each day. There are no pictures in the "snowing" row. That is because it did not snow that month.

When a Picture Is Worth More Than a Piece of Data

Pictographs can be used to show a lot of data. It may not be practical to use one picture for each piece of data. Instead, each picture shows a group. Let us say that Mrs. Lundy's class wants to make a pictograph. They want to show how many people at school choose different lunches. The school has 300 people. So that would be a lot of drawings! Each picture on the graph stands for 20 meals instead of one meal. The key on the graph tells us this. That lets the chart stay much smaller. And it is easier to read.

But, what happens if the total number does not equal a group of 20? Pictographs can use parts of pictures to show smaller amounts. For example, half a picture would be 10 lunches.

But, what if the partial picture is meant to show 9 lunches? Or 11 lunches? It is hard to divide a picture to show this. We cannot be sure of the exact amounts. The person who made the graph must tell us. That is one of the drawbacks of a pictograph.

© Shell Education #50755—Leveled Texts for Mathematics: Data Analysis and Probability

Creating a Pictograph

David wants to show how many families in his school have cats and dogs. He finds out that 208 families have cats. There are 160 families that have dogs. He wants to make a pictograph to show this data.

Step 1: Decide how many items each picture will represent. Often people choose factors of 10. But, different numbers make sense, too. A good choice might be a common factor of your data. David notices that 208 and 160 are both divisible by 16. $208 \div 16 = 13$ and $160 \div 16 = 10$. He chooses to make each picture equal to 16 families. He notes his choice in a key.

Step 2: Decide on a picture. It should be something that relates to your data. Remember that you will need to use it many times. You might choose to make your pictograph by hand. If so, choose something simple. David chooses to use a dog and cat for his pictograph! He labels the rows to show what they represent.

Step 3: Create your graph. Draw pictures to show all the data. Make sure the partial pictures show the right amount. Be neat and clear. Make sure your pictures are the same size. If they are not, it will be hard to see the comparison.

Step 4: Give your graph a descriptive name and key. The title should tell you what the graph shows. The key should show you how much data each picture represents. You can choose to leave out the key only if each picture shows one piece of data.

How Many Families Have Dogs and Cats as Pets	
Key: 1 picture = 16 families	
Dogs	🐕🐕🐕🐕🐕🐕🐕🐕🐕🐕
Cats	🐱🐱🐱🐱🐱🐱🐱🐱🐱🐱🐱🐱🐱

Pictographs in Our Daily Lives

Many newspapers and magazines use pictographs. Pictographs are eye-catching. And they are easy to understand. They can be colorful. And they are an easy way to show comparisons. This makes them an ideal tool. They can help readers understand the data.

You Try It

Create a pictograph. Show the favorite colors of the students in your class.

32

Creating Pictographs

Marvin and Zuna wanted to keep track of the weather in their area for the month of October. Every morning they would get up and check outside. Then they would draw a picture of what the weather was like. The chart they made is below. Why do you think there are no pictures in the "snowing" category?

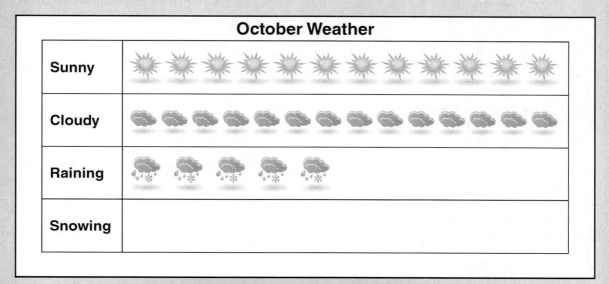

Basic Facts

One way to display data is to use a **pictograph**. A pictograph is a graph that uses pictures to show the data. Sometimes they are called *pictograms* or *picture graphs*. Each picture in the chart can represent one piece of data or a group of data points. In the example above, the children used a picture to represent each day. There are no pictures in the "snowing" row because it did not snow in October.

When a Picture Is Worth More Than a Piece of Data

Often pictographs are used to show large groups of information. It may not be practical to use one picture for each piece of data. Instead, each picture shows a group. For instance, let us say that Mrs. Lundy's class wants to make a pictograph showing how many people at the school chose different lunches. The school has 300 people, so that would be a lot of drawings! Instead of each picture standing for one meal, the key on the graph tells us that each picture stands for 20 meals. That lets the chart stay much smaller and easier to read.

But, what happens if the total number does not equal a group of 20? Pictographs use partial pictures to show smaller amounts. For example, half a picture would equal 10 lunches.

But, what if the partial picture is meant to represent 9 or 11 lunches? It is hard to divide a picture accurately to show this. We cannot be sure of the exact amounts unless the person who made the graph tells us. That is one of the drawbacks to using a pictograph.

33

Creating a Pictograph

David wants to show how many families in his school have cats and dogs. He finds out that 208 families have cats. There are 160 families that have dogs. He decides to make a pictograph to show this data.

Step 1: Decide how many items each picture will represent. Often people choose factors of 10. But sometimes different numbers make sense. A good choice might be a common factor of your data numbers. David notices that 208 and 160 are both divisible by 16. 208 ÷ 16 = 13 and 160 ÷ 16 = 10. He chooses to make each picture equal 16 families. He notes his choice in a key.

Step 2: Decide on a picture. It should be something that relates to your data. Remember that you will need many copies of it. So, if you are making your pictograph by hand, choose something simple. David decides a dog and cat make the most sense for his pictograph! He labels the rows to show what they represent.

Step 3: Create your graph. Fill in enough pictures to represent all the data. If you have partial groups, make sure the partial pictures show the right amount. Be neat and clear. Make sure your pictures are the same size. If they are not, it will be hard to see the comparison.

Step 4: Give your graph a descriptive name and key. The title should tell you exactly what the graph shows. The key should show you how much data each picture represents. You can choose to leave out the key only if each picture shows exactly one piece of data.

How Many Families Have Dogs and Cats as Pets		
Key: 1 picture = 16 families		
Dogs	🐶🐶🐶🐶🐶🐶🐶🐶🐶🐶	
Cats	🐱🐱🐱🐱🐱🐱🐱🐱🐱🐱🐱🐱🐱	

Pictographs in Our Daily Lives

Many newspapers and magazines use pictographs. Pictographs are eye-catching and easy to understand. They can be colorful. And they are an easy way to show comparisons. These qualities make them an ideal tool to help readers better understand data in an article.

You Try It

Create a pictograph showing the favorite colors of the students in your class.

Creating Pictographs

Marvin and Zuna wanted to keep track of the weather in their area for the month of October. Every morning they would get up and check the weather outside. They would draw a picture of what the weather was like and then record their results on the chart below. Why do you think there are no pictures in the "snowing" category?

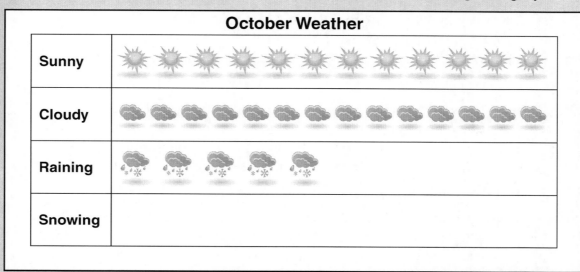

Basic Facts

One way to display data is to use a **pictograph**. A pictograph is a graph that uses pictures to represent the data. Sometimes pictographs are called *pictograms* or *picture graphs*. Each picture in the chart can represent one piece of data or a group of data points. In the example above, the children used a picture to represent each day. There are no pictures in the "snowing" category because it did not snow in October.

When a Picture Is Worth More Than a Piece of Data

Often pictographs are used to display large groups of information. Sometimes it may not be practical to use one picture for each piece of data. Instead, each picture represents a group of data. For instance, let us say that Mrs. Lundy's class wants to make a pictograph showing how many people at the school chose different lunches. The school has 300 people, so that would be a lot of drawings! Instead of each picture representing one meal, the key on the graph tells us that each picture represents 20 meals. This way, the pictograph is smaller and much easier to read.

But, what happens if the total number does not equal a group of 20? Pictographs can also use partial pictures to show smaller amounts. For example, half a picture would equal 10 lunches.

But, what if the partial picture is meant to represent 9 or 11 lunches? It is hard to divide a picture accurately to show this. One of the drawbacks of using a pictograph is that we cannot be sure of the exact amounts unless the person who made the graph informs us.

35

Creating a Pictograph

David wants to show how many families in his school have cats and dogs. He finds out that 208 families have cats and 160 families have dogs. He decides to make a pictograph to represent this data.

Step 1: Decide how many items each picture will represent. Often people choose factors of 10. However, sometimes using different numbers makes more sense. A good choice might be to use a common factor of your data numbers. David notices that 208 and 160 are both divisible by 16. $208 \div 16 = 13$ and $160 \div 16 = 10$. He chooses to make each picture equal to 16 families. He notes his choice in a key on the pictograph.

Step 2: Decide on a picture. Your picture should be something that relates to your data. Remember you will need to be able to re-create it several times. So, if you are making your pictograph by hand, choose something simple. David decides that a dog and cat make the most sense for his pictograph! He labels the rows in the pictograph to show what they represent.

Step 3: Create your graph. Fill in enough pictures to represent all the data. If you have partial groups, make sure the partial pictures show the right amount. Remember to be neat and clear, and make sure your pictures are the same size. If they are not, it will be hard to see the comparison.

Step 4: Give your graph a descriptive name and key. The title should tell you exactly what the graph shows, and the key should show you how much data each picture represents. You can choose to leave out the key only if each picture shows exactly one piece of data.

How Many Families Have Dogs and Cats as Pets	
Key: 1 picture = 16 families	
Dogs	🐕🐕🐕🐕🐕🐕🐕🐕🐕🐕
Cats	🐈🐈🐈🐈🐈🐈🐈🐈🐈🐈🐈🐈🐈

Pictographs in Our Daily Lives

Many newspapers and magazines use pictographs because they are eye-catching, colorful, and easy to understand. They can show comparisons. Pictographs are ideal tools to help readers better understand data in an article.

You Try It

Create a pictograph showing the favorite colors of the students in your class.

Analyzing Pictographs

Libby is selling candy. She wants to raise money. The money is for her Wilderness Rangers troop. This pictograph shows the kinds of candy she sold. Which kind has she sold the least? Which kind has she sold the most? Which two did she sell the same? How many total packages did she sell?

Basic Facts

The problem has four parts. Look at each part.

Which kind of candy has Libby sold the least?

Pictographs make it easy to compare. We look at the graph. We want to find the smallest amount. We use the key. Then we find the exact amounts. Then we compare. This time we can see the lowest amount. It is for taffy. There is only one picture. All the others show more. We know Libby has sold taffy the least.

Which kind of candy has Libby sold the most?

Chocolate bars and lollipops both have three pictures. On lollipops, the last picture is only $\frac{1}{2}$. Libby sold more chocolate bars than lollipops!

Which two tie?

We must compare. We need to find the two that are the same. We do not need the exact amounts. We can look at the chart. "Fruit chews" and "caramels" each have two pictures. The pictograph shows us!

© Shell Education   #50755—Leveled Texts for Mathematics: Data Analysis and Probability

Basic Facts (cont.)

What is the total number sold?

Now we need the exact number. We want the total sold. We need do two things. We will see how many of each type were sold. Then we will add the amounts together. We will know the sum. The key helps. It shows that each full picture stands for 6 packages of candy. Each half picture stands for 3 packages. Our solution looks like this:

Step 1: Find how many of each type is shown on the graph. Use the key. Multiply the number of pictures by the number that each picture stands for.

Chocolate Bars		$3 \times 6 = 18$
Fruit Chews		$2 \times 6 = 12$
Lollipops		$(2 \times 6) + 3 = 15$
Taffy		$1 \times 6 = 6$
Caramels		$2 \times 6 = 12$

Step 2: Find the sum of the products.

Add the products together. We get: $18 + 12 + 15 + 6 + 12 = 63$

Libby has sold 63 packages!

Analyzing Pictographs in Our Daily Lives

You can find pictographs in many places. In the example, they are used to show a candy sale. They can also be used in magazines and newspapers. They are used in ads to compare products. And they are even used on packages of food to show nutrition facts. You need to be able to read the pictographs. That will help you understand the data being shown.

You Try It

Find the total number of fish in the aquarium.

Jeanie's Aquarium	
 = 8 fish	
Guppies	
Tetras	
Goldfish	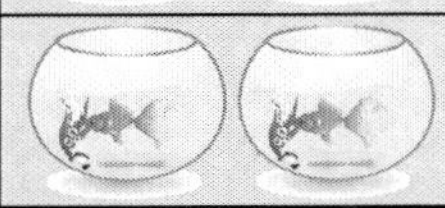

Analyzing Pictographs

Candy Sales

🍭 = 6 packages sold

Chocolate Bars	Fruit Chews	Lollipops	Taffy	Caramels

Libby is having a candy sale. She wants to raise money for her Wilderness Rangers troop. This pictograph shows each kind of candy she sold. Which kind of candy has Libby sold the least? Which candy has she sold the most? Which two candies are in a tie? Can you tell how many packages of candy she has sold in all?

Basic Facts

The problem asks us to find four different pieces of data. Let us look at each question.

Which kind of candy has Libby sold the least?

Pictographs make it easy to compare things. This question asks us to look at the graph. We want to find the smallest amount displayed. We could use the key. Then we could work out the exact amounts and compare them. But in this case, we do not need to do that. We can see that the graph is lowest for taffy. It only shows one picture. All the others show more. We do not need to do any more steps. We know that Libby has sold taffy the least.

Which kind of candy has Libby sold the most?

This question is similar to the first one. You can see that chocolate bars and lollipops both have three pictures. But, on lollipops, the last picture is only $\frac{1}{2}$. That means that Libby sold more chocolate bars than lollipops!

Which two tie?

For this question, we are asked to compare the different columns. We need to find the two that are the same. Again, we do not need the exact amount. We can simply look at the chart and see that the columns for "fruit chews" and "caramels" each have two pictures in them. The pictograph does all the work for us!

© Shell Education   #50755—Leveled Texts for Mathematics: Data Analysis and Probability

Basic Facts (cont.)

What is the total number sold?

Here, we need to find an exact number. To find the total packages of candy sold, we need to figure out how many of each type were sold. Then, we will add them all together to find the sum. We know from the key that each full picture is worth 6 packages of candy. That means that each half picture is worth 3 packages of candy. So, our solution looks like this:

Step 1: Find how many of each type is shown on the graph. To do this, use the key. Multiply the number of pictures by the number that each picture represents.

Chocolate Bars	◎ ◎ ◎	$3 \times 6 = 18$
Fruit Chews	◎ ◎	$2 \times 6 = 12$
Lollipops	◎ ◎ ◐	$(2 \times 6) + 3 = 15$
Taffy	◎	$1 \times 6 = 6$
Caramels	◎ ◎	$2 \times 6 = 12$

Step 2: Find the sum of the products.

Adding the products together, we get: $18 + 12 + 15 + 6 + 12 = 63$
Libby has sold 63 packages of candy so far!

Analyzing Pictographs in Our Daily Lives

Pictographs can be found all over. In the first example, they are used to show a candy sale. They can also be used to show data in magazines and newspapers. They are used in ads to compare products. And they are even used on packages of food to show nutritional content. If you want to understand the data being shown, you need to be able to read the pictographs.

You Try It

Find the total number of fish in the aquarium.

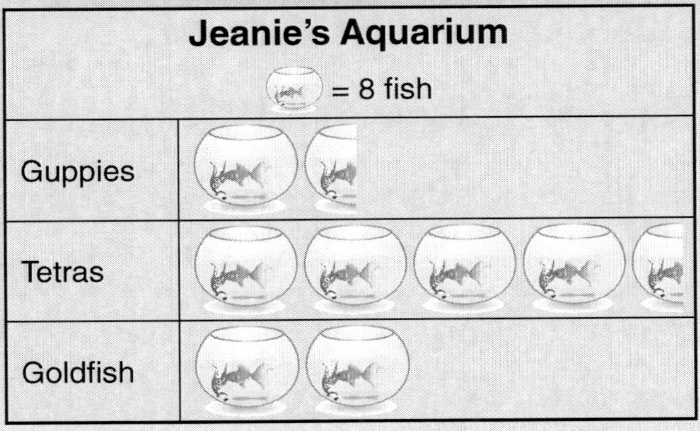

Jeanie's Aquarium		
	🐟 = 8 fish	
Guppies	🐠 🐠	
Tetras	🐠 🐠 🐠 🐠 🐠	
Goldfish	🐠 🐠	

#50755—Leveled Texts for Mathematics: Data Analysis and Probability   © Shell Education

Analyzing Pictographs

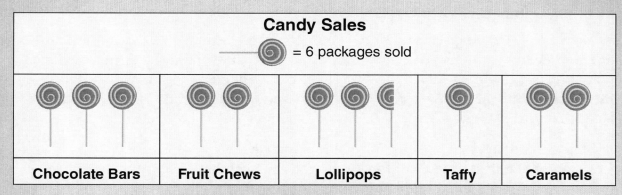

Libby is participating in a candy sale to raise money for her Wilderness Rangers troop. This pictograph shows how many candies she has sold of each type. Which kind of candy has Libby sold the least? Which candy has been her best seller? There are two candy types currently in a tie—which candies are those? Can you tell the total number of candy packages she has sold?

Basic Facts

The problem asks us to find four different pieces of information. Let us examine each question.

Which kind of candy has Libby sold the least?

Pictographs make it easy to compare things. This question asks us to look at the graph and find the smallest amount displayed. We could use the key, work out the exact amounts, and then compare them. But in this case, we do not need to do that. We can easily see that the smallest amount on the graph is for the taffy. It only shows one picture. All the others show more. We do not need to do any more steps to know that Libby has sold taffy the least.

Which kind of candy has Libby sold the most?

This question is very similar to the first one. You can see that chocolate bars and lollipops both have three pictures. But, on lollipops, the last picture is only $\frac{1}{2}$. That means that Libby sold more chocolate bars than lollipops!

Which two tie?

For this question, we are asked to compare the different columns and find the two that are the same. Again, it is not necessary to figure out exact amounts. We can simply look at the chart and see that the columns for "fruit chews" and "caramels" each have two pictures in them. The pictograph does all the work for us!

 #50755—*Leveled Texts for Mathematics: Data Analysis and Probability*

Basic Facts (cont.)

What is the total number sold?

Now, finally we are given a question that needs a specific number as its answer. To discover the total packages of candy sold, we need to figure out how many of each type were sold, and then we will add them together to find the sum. We know from the key that each full picture is worth 6 packages of candy. That means that each half picture is worth 3 packages of candy. So, our solution looks like this:

Step 1: Find how many of each type is displayed on the graph. To do this, use the key. Multiply the number of pictures by the number that each picture represents.

Chocolate Bars	◎ ◎ ◎	$3 \times 6 = 18$
Fruit Chews	◎ ◎	$2 \times 6 = 12$
Lollipops	◎ ◎ ◎	$(2 \times 6) + 3 = 15$
Taffy	◎	$1 \times 6 = 6$
Caramels	◎ ◎	$2 \times 6 = 12$

Step 2: Find the sum of the products.

Adding the products together, we get: $18 + 12 + 15 + 6 + 12 = 63$
Libby has sold 63 packages of candy so far!

Analyzing Pictographs in Our Daily Lives

Pictographs can be found all over. As in our first example, they are used in fundraisers to illustrate progress. They are used in magazines and newspapers to display data in articles, and in advertisements to compare products. They are even used on packages of food to show nutritional content! If you want to be able to understand the information being shown, you need to be able to read and analyze the pictographs.

You Try It

Find the total number of fish in the aquarium.

Jeanie's Aquarium		
	🐟 = 8 fish	
Guppies	🐟 🐟	
Tetras	🐟 🐟 🐟 🐟 🐟	
Goldfish	🐟 🐟	

#50755—Leveled Texts for Mathematics: Data Analysis and Probability   © Shell Education

Analyzing Pictographs

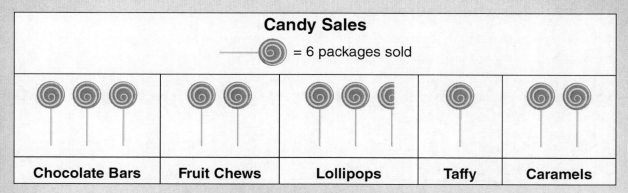

Libby is participating in a candy sale to raise money for her Wilderness Rangers troop. This pictograph shows how many candies she has sold in each of the five categories. Which type of candy has Libby sold the least? Which candy has been her best seller? There are two candy types currently in a tie—which candies are those? Can you determine the total number of candy packages sold?

Basic Facts

The problem asks us to investigate the pictograph in order to find four different pieces of information. Let us examine each question separately in order to find the solutions.

Which kind of candy has Libby sold the least?

Pictographs allow you to easily show a comparison of different things. This question asks us to look at the pictograph and find the smallest amount displayed. We could use the key, work out the exact amounts, and then compare them. However, in this case we do not need to do that. We can easily see that the graph is lowest at the point where the taffy is displayed because it only shows one picture, but all the other categories contain more pictures. No more steps are needed in order to find out that Libby has sold the least amount of taffy.

Which kind of candy has Libby sold the most?

This question is very similar to the first one. You can see that chocolate bars and lollipops both have three pictures. But in the lollipop category, one of the pictures is only $\frac{1}{2}$, which means that Libby sold more chocolate bars than lollipops!

Which two tie?

For this question, we are asked to compare the different columns and find the two that are the same. Again, it is not necessary to figure out the exact amounts in each column. We can simply look at the pictograph and see that the columns for "fruit chews" and "caramels" each have two pictures in them. The pictograph does all the work for us!

43

Basic Facts *(cont.)*

What is the total number sold?

Finally, we are given a question that needs a specific number as its answer. To discover exactly how many packages of candy were sold, we need to figure out how many of each type were sold, and then we will add them all together to find the sum. We know from the key that each full picture is worth 6 packages of candy, which means that each half picture is worth 3 packages of candy. So, our solution looks like this:

Step 1: Find how many of each type is displayed on the graph. To do this, use the key. Multiply the number of pictures by the number that each picture represents.

Chocolate Bars	◎ ◎ ◎	$3 \times 6 = 18$
Fruit Chews	◎ ◎	$2 \times 6 = 12$
Lollipops	◎ ◎ ◎	$(2 \times 6) + 3 = 15$
Taffy	◎	$1 \times 6 = 6$
Caramels	◎ ◎	$2 \times 6 = 12$

Step 2: Find the sum of the products.

Adding the products together, we get: $18 + 12 + 15 + 6 + 12 = 63$
Libby has sold 63 candies so far!

Analyzing Pictographs in Our Daily Lives

Pictographs can be found everywhere. As in our first example, they are used in fundraisers to illustrate progress. They are used in magazines and newspapers to display data in articles, in advertisements to compare products, and even on packages of food to show nutritional content! If you want to be able to understand the information being shown, you need to be able to read and analyze the pictographs.

You Try It

Find the total number of fish in the aquarium.

Jeanie's Aquarium	
= 8 fish	
Guppies	🐟 🐟
Tetras	🐟 🐟 🐟 🐟 🐟
Goldfish	🐟 🐟

#50755—*Leveled Texts for Mathematics: Data Analysis and Probability*   © *Shell Education*

Creating Bar Graphs

Look at the two graphs below. What do you notice?

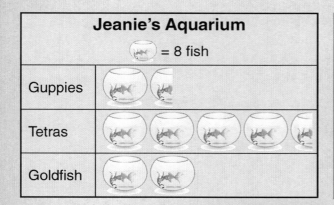

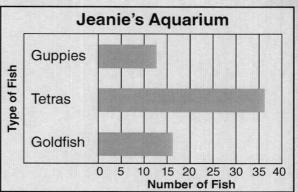

Both graphs show the same data! The graph on the left is a pictograph. The graph on the right is a **bar graph**. Any data can be shown in many ways. The way you choose is up to you!

Basic Facts

A bar graph is like a pictograph. But, it uses bars to show data. Bar graphs can be vertical or horizontal. That means they can go up or across. People may call them column charts or vertical bar graphs. That is when the bars are vertical. Or, people may call them horizontal bar graphs if the bars are horizontal.

The axes (plural of *axis*) are special vertical and horizontal lines on the graph. They are used to label the data shown on the graph. Grid lines are incremental marks. These marks are on the axis of a graph. They show the numeric value for the data in the graph. Look at the bar graph above. The grid lines go from 0 to 40. These marks are in increments of 5.

Making a Bar Graph

Bar graphs are simple to make. Just follow a few steps. Bar graphs need to be neat. And they must be clearly labeled. Let's say Jordan's soccer team wanted to have a party at the end of the season. They took a vote. There were five players who voted for pizza. And seven players voted for a luau. Here is how to make a bar graph of the results.

Step 1: Choose what way your graph will run. Draw the axes.

Step 2: Make grid marks along one axis. Start at 0 if possible. And use equal spaces. We know the largest number we have is 7. So the graph stops at 10. The grid marks are in intervals of 2.

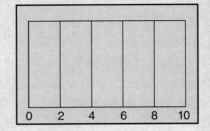

Step 3: Draw bars for each data set. Make them the same thickness. Think about how to space them. Label each bar.

Step 4: Give your graph a title. Label the axes. The finished graph is below.

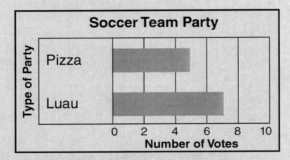

Bar Graphs in Our Daily Lives

Bar graphs are used in business. They are good for showing end results. Stores may use them to show how many items are sold. Or, bar graphs can be used to show the data from surveys. So let's say that a company wants to know if people like their new bicycle product. First they might do a survey. They would ask people some questions. Then they would create a bar graph of the results. They could show the graph to the people in charge of making the new bike. This would help them know how they were doing!

Sample question: Is this bike comfortable to ride? Check one.

Yes   No

You Try It

In one week, Mrs. Jillians collected work from her class. She collected the following assignments: 25 writing, 76 math, 50 social studies, and 48 science. Make a bar graph that shows this data.

#50755—*Leveled Texts for Mathematics: Data Analysis and Probability*   © *Shell Education*

Creating Bar Graphs

Look at the two graphs below. What do you notice?

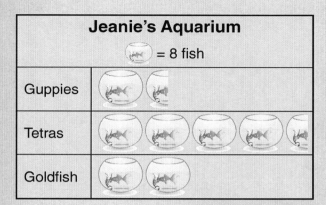

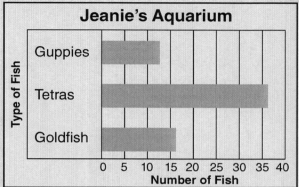

Both graphs show the same data! The graph on the left is a pictograph. The graph on the right is a **bar graph** that shows the same data. Any data can be shown in a variety of ways. The way you choose is up to you!

Basic Facts

A bar graph is like a pictograph. But, it uses bars instead of pictures to show data. Bar graphs can be vertical or horizontal. People may call them column charts or vertical bar graphs. That is when the bars are vertical. Or, people may call them horizontal bar graphs if the bars are horizontal.

The axes (plural of *axis*) are special vertical and horizontal lines on the graph. They are used to label the data shown on the graph. Grid lines are incremental marks. These marks are on the axis of a graph. They show the numeric value for the data in the graph. On the bar graph above, there are grid lines showing the values from 0 to 40. These marks are in increments of 5.

Making a Bar Graph

Bar graphs are simple to make. Just follow a few steps. Bar graphs need to be neat. And the information must be clearly labeled. For example: Jordan's soccer team wanted to have a party at the end of the season. They took a vote. There were five players who voted for pizza. And seven players voted for a luau. Here is how to make a bar graph of the results.

Step 1: Decide what direction your graph will run. Draw the axes.

Step 2: Make grid marks along one axis for reference. Start at 0 if possible. And use equal intervals. We know the largest number we have is 7. So the graph stops at 10, with grid marks at intervals of 2.

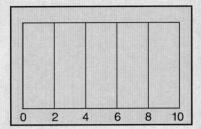

Step 3: Draw bars for each data set. Make them the same thickness. Think about how to space them from bottom to top. Label each bar.

Step 4: Give your graph a title. Label the axes. The finished graph is below.

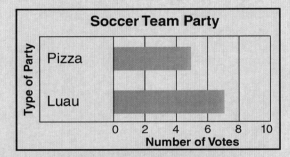

Bar Graphs in Our Daily Lives

Bar graphs are used in business. They are good for showing one-time results. Stores may use them to show inventory or sales by item. Or, bar graphs can be used to show the data from surveys. So let's say that a company wants to know if people like their new bicycle product. First their marketing department might do a survey. They would ask their customers some questions. Then they would create a bar graph of the results. They could show the graph to the people in charge of making the new bike. This would help them know how they were doing!

Sample question: Is this bike comfortable to ride? Check one.

Yes No

You Try It

One week Mrs. Jillians collected 25 writing assignments, 76 math assignments, 50 social studies assignments, and 48 science assignments from the students in her class. Make a bar graph that shows this data.

Creating Bar Graphs

Look at the two graphs below. What do you notice?

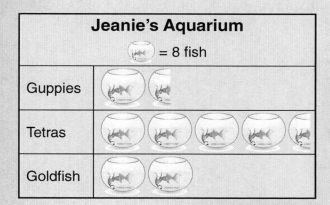

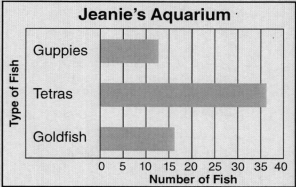

Both graphs show the same information! The graph on the left is a pictograph, and the graph on the right is a **bar graph** that displays the same data. Any data can be shown in a variety of ways. The way you choose is up to you!

Basic Facts

A bar graph is like a pictograph. But, it uses bars instead of pictures to display data. Bar graphs can be vertical or horizontal. When the bars are vertical, people may call the graphs column charts, vertical bar graphs, or just bar graphs. When the bars are horizontal, people may call the graphs horizontal bar graphs or just bar graphs.

The axes (plural of *axis*) are the vertical and horizontal lines that are used to label the information displayed on a graph. Grid lines are incremental marks on the axis of a graph. They show the numeric value for the data in the graph. On the bar graph above, there are grid lines showing the values from 0 to 40 in increments of 5.

Making a Bar Graph

Bar graphs are simple to construct. Just follow a few steps. Bar graphs need to be neat and the information must be clearly labeled. Here is an example: Jordan's soccer team voted on which kind of party they wanted to have to celebrate the end of the season. There were five players who voted for pizza and seven who voted for a luau. Here is how to make a bar graph of the results.

Step 1: Decide what direction your graph will run, and draw the axes.

Step 2: Make grid marks along one axis for reference. Start at 0 if possible, and use equal intervals. We know the largest number we have is 7, so the graph stops at 10, with grid marks at intervals of 2.

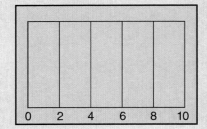

Step 3: Draw bars of equal thickness for each data set. Make sure they are spaced equally from top to bottom. Label each bar.

Step 4: Give your graph a title and label the axes. The finished graph is below.

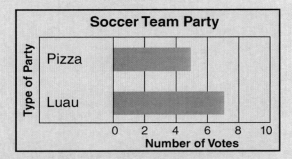

Bar Graphs in Our Daily Lives

Bar graphs are often used in business. They are very good at displaying one-time results. Stores may use them to show inventory or sales by item, for instance. Or, bar graphs can be used to display the data from surveys. So, let's say that a company wants to know how people like their new bicycle product. First their marketing department might do a survey to ask the customers. Then they would create a bar graph of the results. They could show the graph to the people in charge of making the new bike so they could know how they were doing!

Sample question: Is this bike comfortable to ride? Check one.

☐ Yes ☐ No

You Try It

One week Mrs. Jillians collected 25 writing assignments, 76 math assignments, 50 social studies assignments, and 48 science assignments from the students in her class. Make a bar graph that shows this data.

Creating Bar Graphs

Investigate the two graphs displayed below. What do you notice about the graphs?

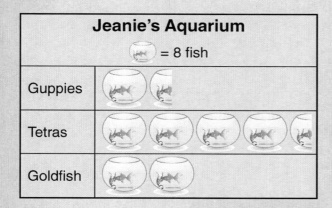

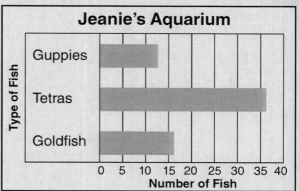

Both graphs display the same information! The graph on the left is a pictograph and the graph on the right is a **bar graph** that represents the same data. Any data can be displayed in many different ways, but the way you choose to display it is entirely up to you!

Basic Facts

A bar graph is like a pictograph, but it uses bars instead of pictures to display information. Bar graphs can be vertical or horizontal. When the bars are vertical, people may call the graphs column charts, vertical bar graphs, or just bar graphs. When the bars are horizontal, people may call the graphs horizontal bar graphs or just bar graphs.

The axes (plural of *axis*) are the vertical and horizontal lines that are used to label the information displayed on a graph. Grid lines are incremental marks on the axis of a graph that show the numeric value for the data in the graph. On the bar graph above, there are grid lines showing the values from 0 to 40 in increments of 5.

51

Making a Bar Graph

Bar graphs are simple and straightforward to construct—just follow a few steps! Bar graphs need to be neat and the information must be organized and clearly labeled. Here is an example: Jordan's soccer team voted on which kind of party they wanted to have to celebrate the end of the season. There were five players who voted for pizza and seven who voted for a luau. Here is a list of the steps to follow to make a bar graph of the results of the vote.

Step 1: Decide what direction your graph will run, and draw the axes.

Step 2: Make grid marks along one axis for reference. Start at 0 if possible, and use equal intervals. We know the largest number we have is 7, so the graph stops at 10, with grid marks at intervals of 2.

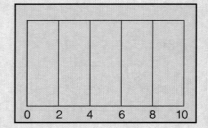

Step 3: Draw bars of equal thickness for each data set and be sure to space them equally from top to bottom. Don't forget to label each bar correctly.

Step 4: Title your graph and label the axes. The finished graph is displayed below.

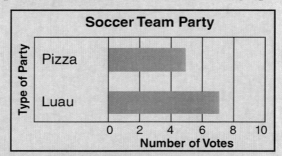

Bar Graphs in Our Daily Lives

Businesses use bar graphs to organize information and to display one-time results. Stores may use them to display inventory or to show sales by item. Bar graphs can also be used to display the information taken from surveys. If a company wants to know how people like their new bicycle product, their marketing department might conduct a survey to ask customers questions about the product. Then they would create a bar graph of the results. They could present the graph to the people making the new bike so they could know how they were doing!

Sample question: Is this bike comfortable to ride? Check one.

Yes   No

You Try It

One week Mrs. Jillians collected 25 writing assignments, 76 math assignments, 50 social studies assignments, and 48 science assignments from the students in her class. Make a bar graph that displays this data.

52

Analyzing Bar Graphs

Mina's class made a bar graph. It showed the kinds of books read by the students for one month. Which type, or genre, of book was most popular? Which kinds of books were read fewer than 20 times? How many science fiction books were read that month?

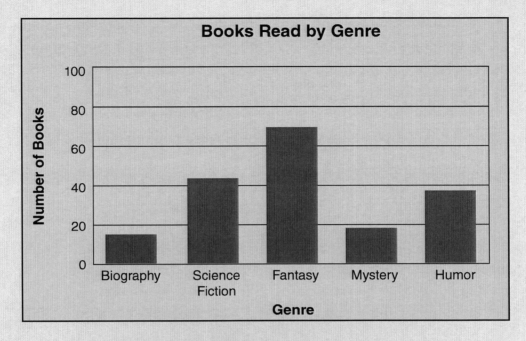

Basic Facts

We need to find three pieces of data. Look at each question.

Which genre was the most popular?

Bar graphs show extremes. Compare the height of the bars. We can see what is similar. We can see what is different. The bar for fantasy books goes higher than the others. More fantasy books were read that month.

Which genres were read fewer than 20 times?

This question is like the first one. We need to see which bars are less than 20. We need to look at all the bars. There are two bars below that line. Mystery books were read less than 20 times that month. So were biography books.

© Shell Education   #50755—Leveled Texts for Mathematics: Data Analysis and Probability

Basic Facts (cont.)

About how many science fiction books were read that month?

Now we make an estimate. We use the scale. The scale does not show single digit numbers. Think about lines splitting up the rows.

Step 1: Picture a line that splits the section in half.

Look at the lines for 40 and 60 books. Think of a line splitting that section into two parts. The parts will be equal. That new line would be at 50 books. The "science fiction" bar barely goes past the line for 40. The bar does not come up to the 50 line. So we know that there are less than 50 science fiction books.

Step 2: Think about more lines splitting the new section into more parts.

Try to picture splitting the 40 to 50 section. You can make 5 equal parts. Each of those parts would stand for 2 books.

Step 3: Compare the bar to the new grid lines.

The bar looks like it comes up to about the first of those 5 lines. So, 42 is a good estimate of how many science fiction books were read that month.

Analyzing Bar Graphs in Our Daily Lives

Bar graphs are common. They are easy to make. They are easy to read. They help us compare sets of data. They are used in all forms of media. You will see bar graphs in newspapers and magazines. You can also find them on TV and on computers. Some video games use bar graphs to show scores!

You Try It

Look at the bar graph on the last page. Estimate how many books of each type Mina's class read over the course of the month. Fill in the data table below. Science Fiction is already filled in for you.

Biography	Science Fiction	Fantasy	Mystery	Humor
	42			

#50755—*Leveled Texts for Mathematics: Data Analysis and Probability*   © Shell Education

Analyzing Bar Graphs

Mina's class made a bar graph. It showed the kinds of books read by the students for one month. Which type, or genre, of book was the most popular? Which genres were read fewer than 20 times? How many science fiction books were read that month?

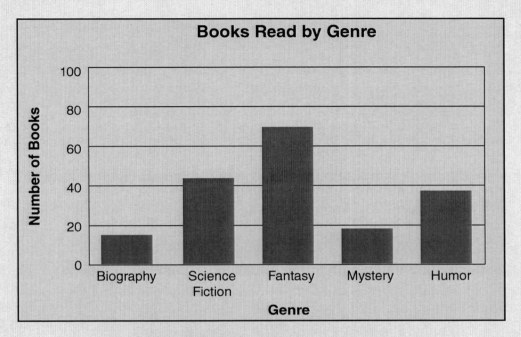

Books Read by Genre

Basic Facts

The problem asks us to find three pieces of data. Let us look at each question.

Which genre was the most popular?

Bar graphs are good at showing extremes. We can compare the height of the bars. This lets us see the differences and similarities of the data set. Look at the height of the bars in this graph. We can see that the bar for fantasy books goes much higher than any of the others. Thus we can tell that fantasy books were the most popular that month.

Which genres were read fewer than 20 times?

This question is like the first one. But, now we are asked to compare bars to a certain point on the scale. So we need to look at the bars that do not go above the line showing 20 books. There are two bars that stay below that line. Mystery and biography books were each read less than 20 times that month.

© Shell Education　　　#50755—Leveled Texts for Mathematics: Data Analysis and Probability

Basic Facts (cont.)

About how many science fiction books were read that month?

Now we must make an estimate based off the scale. The scale does not show single digit numbers. So, the best way is to imagine lines splitting each row into smaller sections.

Step 1: Imagine a line dividing the key section in half.

Look at the section between 40 and 60 books. Think of a line splitting the 40–60 section into two parts. The parts will be equal. That line would be at 50 books. The science fiction bar barely goes past the line for 40. The bar does not come up to the 50 line. So, now we know that there are fewer than 50 books in the science fiction data set.

Step 2: Imagine more lines dividing the new section into workable parts.

Imagine splitting the 40 to 50 section into 5 parts. The parts will be equal. Each of those parts would represent 2 books.

Step 3: Compare the bar to the new grid lines.

The bar looks like it comes up to about the first of those 5 lines. So, 42 might be a good estimate of how many science fiction books were read that month.

Analyzing Bar Graphs in Our Daily Lives

Bar graphs are common. They are easy to make. And they are easy to read. They help us see a comparison between sets of data. Because of these strengths, they are used in all forms of media. You will see bar graphs in newspapers and magazines. You can also find them on TV and on computers. Some video games use bar graphs to show scores!

You Try It

Look at the bar graph on the previous page. Estimate how many books of each type Mina's class read over the course of the month. Fill in the data table below with your answers. Science fiction is already filled in for you.

Biography	Science Fiction	Fantasy	Mystery	Humor
	42			

#50755—Leveled Texts for Mathematics: Data Analysis and Probability

Analyzing Bar Graphs

Mina's class made a bar graph showing the kinds of books read by the students for one month. Which type, or genre, of book was the most popular? Which genres were read fewer than 20 times? About how many science fiction books were read that month?

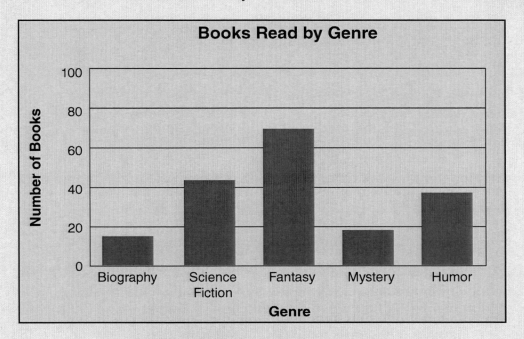

Basic Facts

The problem asks us to find three pieces of information. Let us examine each question.

Which genre was the most popular?

Bar graphs are particularly good at displaying extremes. Comparing the height of the bars allows us to see differences and similarities in the data set. If we look at the height of the bars in this graph, we can see that the bar representing fantasy books goes significantly higher than any of the others. Thus we can tell easily from a glance that fantasy books were the most popular that month.

Which genres were read fewer than 20 times?

This question is very similar to the first one. But in this case, we are asked to compare bars to a specific point on the scale. To find the answer, we need to look at the bars that do not rise above the line designating 20 books. There are two bars that stay below that line. Mystery and biography books were each read fewer than 20 times that month.

© Shell Education   #50755—Leveled Texts for Mathematics: Data Analysis and Probability

Basic Facts *(cont.)*

About how many science fiction books were read that month?

In this question we must make an estimate based off the scale. The scale does not show single digit numbers. So, the best approach is to imagine lines dividing each row into smaller sections.

Step 1: Imagine a line dividing the key section in half.

Look at the section between 40 and 60 books. Think of a line dividing the 40 to 60 section into two equal parts. That line would be at 50 books. The science fiction bar barely reaches past the line for 40. The bar does not come up to that line. So, now we know that there are fewer than 50 books in the science fiction data set.

Step 2: Imagine more lines dividing the new section into workable parts.

Imagine dividing the 40 to 50 section into 5 equal parts. Each of those parts would represent 2 books.

Step 3: Compare the bar to the new, imagined grid lines.

The bar looks like it comes up to about the first of those 5 lines. So, 42 might be a good estimate of how many science fiction books were read that month.

Analyzing Bar Graphs in Our Daily Lives

Bar graphs are very common. They are easy to make and easy to read. They are very useful for letting people see a quick comparison between sets of data. Because of these strengths, they are used frequently in all forms of media. You will see bar graphs in newspapers, in magazines, on television, and in computer programs. Many video games even use bar graphs to display scores!

You Try It

Look at the bar graph on the previous page. Estimate how many books of each type Mina's class read over the course of the month. Fill in the data table below with your answers. Science Fiction is already filled in for you.

Biography	Science Fiction	Fantasy	Mystery	Humor
	42			

#50755—*Leveled Texts for Mathematics: Data Analysis and Probability*

Analyzing Bar Graphs

Mina's class created a bar graph displaying the kinds of books read by the students for one month. Which type, or genre, of book was the most popular? Which genres were read fewer than 20 times? About how many science fiction books were read that month?

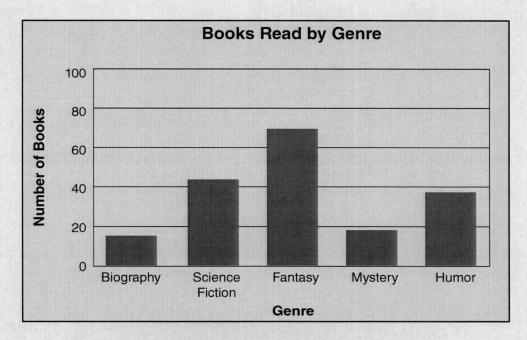

Basic Facts

The problem requires us to find three different pieces of information in order to answer all three questions. Let us examine each question separately.

Which genre was the most popular?

Bar graphs are particularly good at displaying extremes. Comparing the heights of the bars allows us to discern the differences and similarities in the data set. If we look at the heights of the bars in this graph, we can see that the bar representing fantasy books rises significantly higher than any of the others. Thus we can tell easily from a glance that fantasy books were the most widely-read that month.

Which genres were read fewer than 20 times?

This question is very similar to the first one. But in this case, we are being asked to compare the bars to a specific point on the scale. To find the answer, we need to consider the bars that do not rise above the line designating 20 books. There are two bars that stay below that line. Mystery and biography books were each read fewer than 20 times that month.

59

Basic Facts *(cont.)*

About how many science fiction books were read that month?

In this question we must make an estimate based on the scale. The scale does not show single digit numbers. So, the best approach is to imagine lines that divide each row into smaller subsections; that should help us to estimate the correct number

Step 1: Imagine a line dividing the key section in half.

Look at the section between 40 and 60 books and think of a line dividing the 40 to 60 section into two equal parts. That line would represent 50 books. The science fiction bar reaches barely reaches into the line that signifies 40. So, now we know that there are fewer than 50 books in the science fiction data set.

Step 2: Imagine more lines dividing the new section into workable parts.

Imagine dividing the 40 to 50 section into 5 equal parts. Each of those subsections would represent 2 books.

Step 3: Compare the bar to the new, imagined grid lines.

It appears as though the bar comes up to about the first of those 5 lines, so 42 might be an accurate estimate of how many science fiction books were read that month.

Analyzing Bar Graphs in Our Daily Lives

Bar graphs are used frequently because they are simple to make and easy to read. They are very useful for displaying quick comparisons between sets of data. Because of these strengths, they are commonly used in various forms of media. You will see bar graphs in newspapers, in magazines, on television, and in computer programs. Countless numbers of video games even utilize bar graphs to display scores!

You Try It

Look at the bar graph on the previous page. Estimate how many books of each type Mina's class read over the course of the month and fill in the data table below. Science fiction is already completed for you.

Biography	Science Fiction	Fantasy	Mystery	Humor
	42			

60

Creating Line Graphs

Amal tracks his high scores on a video game. He does this every day. He uses a graph. Look at his graph. Which day was he not able to play?

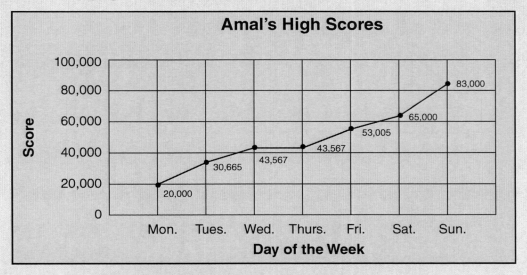

Amal's High Scores

On Thursday Amal's score does not change. That must be the day when he could not play!

Basic Facts

A **line graph** shows data. It shows changes over time. It uses connecting points on a grid. The axes (plural of *axis*) are special horizontal and vertical lines on the graph. These are used to label the graph. Look at the chart above. One axis shows time. The time is shown as days of the week. The vertical axis shows the scores. On many graphs the horizontal axis is called the *x*-axis. The vertical axis is called the *y*-axis.

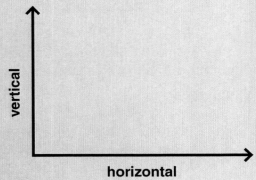

An **ordered pair** is a pair of coordinates that shows a place on a grid. The first coordinate in the pair tells you how many to go over. The second tells you how many to go up. Look at the graph above. The ordered pair (Monday, 20,000) is for the first point. To make a line graph, plot ordered pairs. Then connect them with lines.

Creating a Line Graph

It is easy to create a line graph. You just have to follow the right steps. Look at this example. Every year the city's Parks Department plants trees. In 2006, they planted 40 trees. In 2007, they planted 57 trees. In 2008, they planted 45 trees. In 2009, they planted 60 trees. In 2010, they planted 34 trees. And in 2011, they planted 52 trees. Create a line graph to show this data.

Step 1: Draw and label your axes. First decide what scale to use. Choose the increments to use. Our data ranges from 34 to 60. The scale should be from 0 to 70. Make your scale in groups of ten.

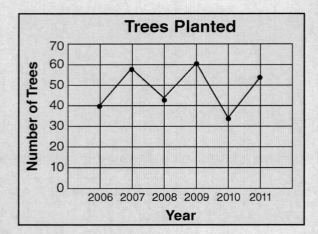

Step 2: Plot your data. Use ordered pairs to plot each point. In this case, the first ordered pair would be (2006, 40).

Step 3: Connect the points.

Step 4: Give your graph a title. The finished graph is shown above.

Line Graphs in Our Daily Lives

Line graphs show change over time. They show trends or patterns. Many companies use line graphs. They show their sales. If sales go up, the line would rise to the right. If sales stay the same, the line would be flat. If sales go down, the line would look like a slide.

You Try It

The aquarium is filling a large tank. They track the water in it. They see the amounts over time. Make a line graph. Use the data below.

Time	8:00 A.M.	12:00 P.M.	4:00 P.M.	8:00 P.M.	12:00 A.M.	4:00 A.M.
Gallons of Water	1,000	9,000	17,000	25,000	33,000	41,000

#50755—*Leveled Texts for Mathematics: Data Analysis and Probability*   © Shell Education

Creating Line Graphs

Amal has been keeping track of his high score on a video game. He did this every day for a week. Can you tell from this graph which day he was not able to play?

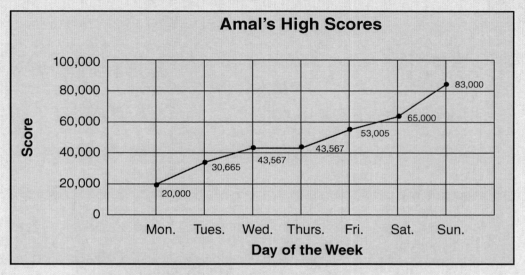

Thursday is the only day Amal's score does not change. So that must be the day when he could not play!

Basic Facts

A **line graph** is a graph that shows how data changes over time. This is shown by connecting points on a grid. The axes (plural of *axis*) are special vertical and horizontal lines on the graph. These are used to label the information. On the chart above, the horizontal axis shows time. The time is shown in the form of days of the week. The vertical axis shows the scores. On many graphs the horizontal axis is called the *x*-axis. The vertical axis is called the *y*-axis.

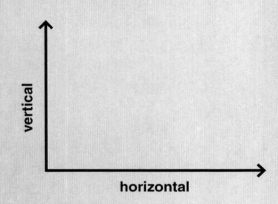

An **ordered pair** is a pair of coordinates that shows a place on a grid. The first coordinate in the pair tells you how many to go over. The second tells you how many to go up. In the graph above, the first point is written with the ordered pair (Monday, 20,000). To make a line graph, plot ordered pairs. Then connect them with lines.

© *Shell Education*   *#50755—Leveled Texts for Mathematics: Data Analysis and Probability*

Creating a Line Graph

It is easy to create a line graph. You just have to follow the right steps. Here is an example: Every year, the city's Parks Department plants trees. In 2006, they planted 40 trees. In 2007, they planted 57 trees. In 2008, they planted 45 trees. In 2009, they planted 60 trees. In 2010, they planted 34 trees. And in 2011, they planted 52 trees. Create a line graph to show this data.

Step 1: Draw and label your axes. For this step you will need to decide what scale to use. And you need to choose what increments to use. Our data ranges from 34 to 60. A good scale would be from 0 to 70. It makes sense to use increments of groups of 10.

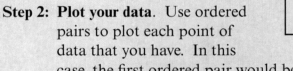

Trees Planted

Step 2: Plot your data. Use ordered pairs to plot each point of data that you have. In this case, the first ordered pair would be (2006, 40).

Step 3: Connect the points.

Step 4: Give your graph a title. The finished graph is shown above.

Line Graphs in Our Daily Lives

Line graphs show change over time. They are useful for showing trends or patterns. Many companies use line graphs. They help them get a quick idea of how their sales are doing. If sales are going up, the line graph will look like a rugged mountainside rising up to the right. If sales are staying the same, the graph will be flat. If sales are going down, the graph will look like a slide.

You Try It

The aquarium is filling a large tank. They keep track of how much water is in it over time. Make a line graph to show the data below.

Time	8:00 A.M.	12:00 P.M.	4:00 P.M.	8:00 P.M.	12:00 A.M.	4:00 A.M.
Gallons of Water	1,000	9,000	17,000	25,000	33,000	41,000

#50755—*Leveled Texts for Mathematics: Data Analysis and Probability*   © *Shell Education*

Creating Line Graphs

Amal has been keeping track of his high score on a video game every day for a week. Can you tell from this graph which day he was not able to play?

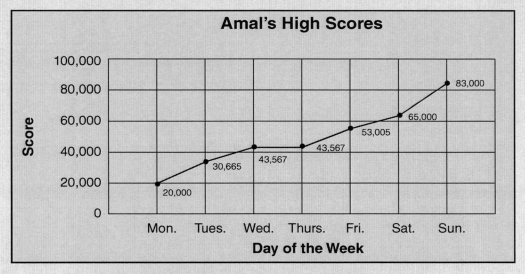

Thursday is the only day Amal's score does not change, so that must be the day when he could not play!

Basic Facts

A **line graph** is a graph that shows how data changes over time by connecting points on a grid. The axes (plural of *axis*) are the vertical and horizontal lines that are used to label the information displayed on a graph. On the chart above, the horizontal axis shows time in the form of days of the week. The vertical axis shows the scores. On many graphs the horizontal axis is called the *x*-axis and the vertical axis is called the *y*-axis.

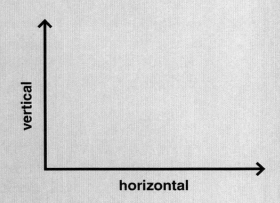

An **ordered pair** is a pair of coordinates that designates a position on a grid. The first coordinate in the pair tells you how many to go over and the second tells you how many to go up. In the graph above, the first point is written with the ordered pair (Monday, 20,000). To create a line graph, plot ordered pairs and then connect them with line segments.

 #50755—Leveled Texts for Mathematics: Data Analysis and Probability

Creating a Line Graph

It is easy to create a line graph if you follow the correct steps. Here is an example: Every year, the city's Parks Department plants trees. In 2006, they planted 40 trees. In 2007, they planted 57. In 2008, they planted 45. In 2009, they planted 60. In 2010, they planted 34 and in 2011, they planted 52. Create a line graph to show this data.

Step 1: Draw and label your axes. For this step you will need to decide what scale and increments to use for your graph. Our data ranges from 34 to 60. A reasonable scale would be from 0 to 70. Increments in groups of 10 make sense.

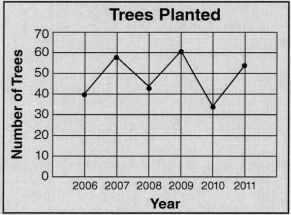

Step 2: Plot your data. Use ordered pairs to plot each point of data that you have. In this case, the first ordered pair would be (2006, 40).

Step 3: Connect the points.

Step 4: Give your graph a title. The finished graph is shown at right.

Line Graphs in Our Daily Lives

Line graphs show how information changes over time. They are very useful for showing trends or patterns. Many companies use line graphs to get a quick idea of how their sales are doing. If sales are going up, a line graph of the data will look like a rugged mountainside rising to the right. If sales are staying the same, the graph will be flat. If sales are declining, the graph will look like a slide.

You Try It

The aquarium is filling a large tank. They keep track of how much water is in it over time. Make a line graph to display the following data.

Time	8:00 A.M.	12:00 P.M.	4:00 P.M.	8:00 P.M.	12:00 A.M.	4:00 A.M.
Gallons of Water	1,000	9,000	17,000	25,000	33,000	41,000

66

Creating Line Graphs

Amal has been recording his high scores on a video game every day for a week. Examine his graph below. Can you determine from the data in this graph which day Amal was not able to access his video game?

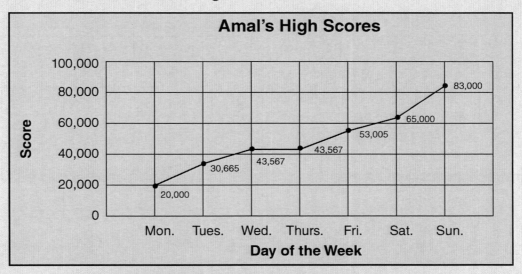

Amal's High Scores

Thursday is the only day in which there is no change in Amal's score, therefore that must be the day when he could not play his video game!

Basic Facts

A **line graph** is a graph that portrays how data changes over time by connecting points on a grid. The axes (plural of *axis*) are the vertical and horizontal lines that are used to label the information displayed on the graph. On the chart above, the horizontal axis shows the time in the form of days of the week and the vertical axis shows the scores in increments of 20,000. On these types of graphs the horizontal axis is called the *x*-axis and the vertical axis is called the *y*-axis

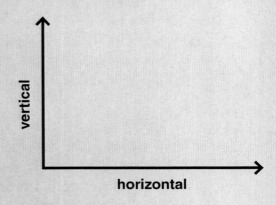

An **ordered pair** is a pair of coordinates that designates the position of data on a grid. The first coordinate in the pair indicates how many spaces to go over and the second coordinate indicates how many spaces to go up. In the graph above, the first point is plotted by the ordered pair (Monday, 20,000). To create a line graph, plot ordered pairs and then connect them with line segments.

67

Creating a Line Graph

It is fairly easy to create a line graph if you follow the correct steps. Here is an example: Every year, the city's Parks Department plants trees. In 2006, they planted 40 trees and in 2007, they planted 57. In 2008, they planted 45 and in 2009, they planted 60. In 2010, they planted 34 and in 2011, they planted 52. Create a line graph to display this data.

Step 1: Draw and label your axes. For this step you will need to decide what scale and increments to use for your graph. Our data ranges from 34 to 60 so a reasonable scale would be from 0 to 70. Increments in groups of 10 make sense.

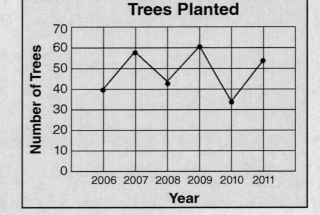

Step 2: Plot your data. Use ordered pairs to plot each point of data that you have. In this example, the first ordered pair would be (2006, 40).

Step 3: Connect the points.

Step 4: Give your graph a title. The finished graph is shown above.

Line Graphs in Our Daily Lives

Line graphs illustrate how information changes over time, and are particularly useful in showing trends, or patterns.. Companies use line graphs to get a quick overview of their sales. If sales rise, a line graph of the data will appear to climb to the right. If sales are staying the same, the graph will be flat. If sales are declining, the graph will resemble a slide.

You Try It

The aquarium is filling a large tank with water. They decide to keep track of how much water is in the tank over time. Make a line graph to display the following data.

Time	8:00 A.M.	12:00 P.M.	4:00 P.M.	8:00 P.M.	12:00 A.M.	4:00 A.M.
Gallons of Water	1,000	9,000	17,000	25,000	33,000	41,000

#50755—*Leveled Texts for Mathematics: Data Analysis and Probability*

Analyzing Line Graphs

Maria has a model rocket club. They meet when the weather is sunny. They launch homemade rockets. This graph shows how many launches they had. It shows each month the club met. In what month did the club launch the most rockets? What is the least number of launches during an active month? What trends do you notice?

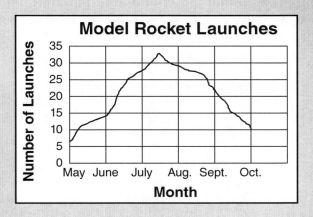

Basic Facts

The problem asks for three pieces of data. Look at each question.

During which month did Maria's rocket club launch the most rockets?

Line graphs show changes over time. They show comparisons. The ticks at the bottom of this graph split it into months. Look at the first section. The data is for the month of May. The space between the second and third ticks is for June. Look at the highest point on the chart. It is between the two marks for July. So, July was the month with the most launches!

What is the fewest number of rockets that the club launched during an active month?

It can be hard to find numbers on a line graph. They are not marked exactly. Look at the first point on the chart. It is hard to tell if it shows 7 or 8 launches.

You could take a ruler and measure. You would see that the point is below halfway between 5 and 10. So it shows 7 launches. That is a lot of work! Luckily, the question asks about the lowest point on the graph. The graph ends on the line for the number 5. The smallest number of launches was 5.

What trends do you notice in the data?

A **trend** is a pattern. It can be seen in a set of data over time. Trends help people make guesses about future data in the same set. Look at the graph above. The trend early on showed the launches increasing. In July, the launches peaked. After that, they went down. Why do you think there were more launches in the middle of the season?

© Shell Education   #50755—Leveled Texts for Mathematics: Data Analysis and Probability

Finding Missing Data on a Line Graph

There is a lot of data that can be shown on line graphs. Sometimes an exact number can be found. And other times you have to use estimation. This graph shows George's height. It goes from the time he was born until he was 10 years old. Find how tall he was when he was $7\frac{1}{2}$ years old.

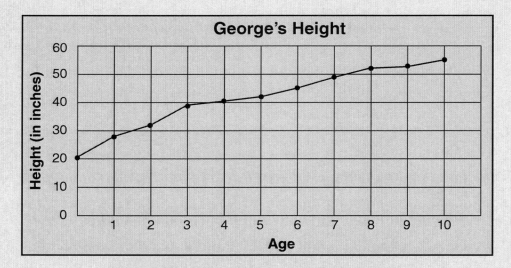

George's Height

Step 1: Find the point on the horizontal axis that is midway between 7 and 8. Use a straight edge to measure up from that point to the place where it intersects with the graph.

Step 2: Use the straight edge again. Find the point on the vertical axis to the left of this point.

When George was $7\frac{1}{2}$ years old, he was 50 inches tall.

Analyzing Line Graphs in Our Daily Lives

Line graphs are good for showing change over time. So, they are often used in finance. Let us say a person owns part of a company. This is called "stock." At times that stock is worth more. Other times it is worth less. People like to make money with their stocks. So they try to buy it when it is cheap. Then they can sell it when it is worth more. A line graph can help them track how their stock is doing. This can help them know when to buy more and when to sell.

You Try It

Use the graph above. Find how tall George was when he was $4\frac{1}{2}$ years old.

Analyzing Line Graphs

Maria's model rocket club meets during sunny weather. They launch the rockets that they make. This graph shows how many rocket launches they had. It shows each month the club met this year. In what month did the club launch the most rockets? What is the fewest number of rockets that the club launched during an active month? What trends do you notice in this data?

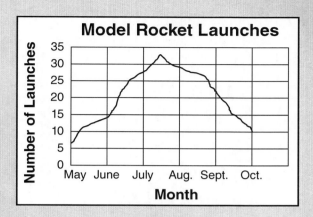

Basic Facts

The problem asks us to find three pieces of data. Let us look at each question.

During which month did Maria's rocket club launch the most rockets?

Line graphs are good at showing changes in data over time. They are good for showing comparisons. The ticks at the bottom of this graph split it into months. From the vertical axis to the first tick, the data for May the first month. The space between the second and third ticks is for June, etc. Let's look at the data on the chart. The highest point takes place in the space between the two marks for the month of July. So, July was the month with the most rocket launches!

What is the fewest number of rockets that the club launched during an active month?

It can be hard to find numbers on a line graph unless they are marked exactly. If you look at the first point on the chart, it is hard to tell if that point shows 7 or 8 launches for the first day of May. You could take a ruler and measure the distance. You would see that the point is below halfway between 5 and 10. So it shows 7 launches. But that is a fair bit of work! Luckily, the question we are asked is about the lowest point on the graph. The graph ends on the line for the number 5. Five is the fewest number of rockets that the club launched during an active month!

What trends do you notice in the data?

A **trend** is a pattern that can be seen in a set of data over time. Trends can help people make guesses about future data in the same set. Look at the graph at the top of this page. The trend in the early months was for the rocket launches to increase. In July, the launches peaked. And, after that, they went down for the rest of the season. Can you think of a reason that fewer rockets might be launched at the beginning and ending of the season than during the middle?

 #50755—Leveled Texts for Mathematics: Data Analysis and Probability

Finding Missing Data on a Line Graph

There is a lot of data that can be shown on line graphs. Sometimes an exact number can be found. And other times you have to use estimation. This graph shows George's height from the time he was born until he was 10 years old. Using the graph, find how tall George was when he was $7\frac{1}{2}$ years old.

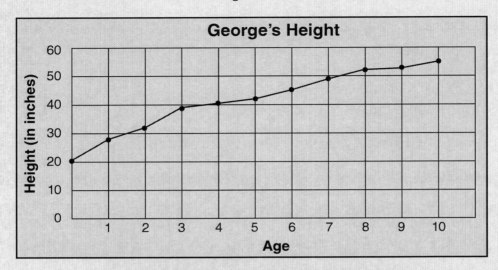

George's Height

Step 1: Find the point on the horizontal axis that is midway between 7 and 8. Use a straight edge to measure up from that point to the place where it intersects with the graph.

Step 2: Using the straight edge, find the point on the vertical axis directly to the left of this point.

When George was $7\frac{1}{2}$ years old, he was 50 inches tall.

Analyzing Line Graphs in Our Daily Lives

Line graphs are good for showing change over time. So they are often used in finance. Let us say a person owns part of a company. This is called "stock." At times that stock is worth different amounts. People like to make money with their stocks. So they try to buy it when it is cheap. Then they can sell it when it is worth more. A line graph can help them track how their stock is doing. This can help them know when to buy more and when to sell.

You Try It

Use the graph above to find how tall George was when he was $4\frac{1}{2}$ years old.

#50755—*Leveled Texts for Mathematics: Data Analysis and Probability*

Analyzing Line Graphs

Maria's model rocket club meets during sunny weather to launch the rockets that they make. This graph shows how many rocket launches they had during each month that the club was active this year. During which month did her club launch the most rockets? What is the fewest number of rockets that the club launched during an active month? What trends do you notice in this data?

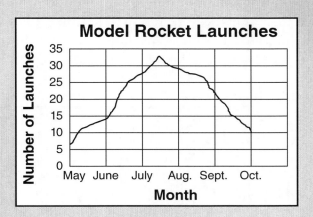

Basic Facts

The problem asks us to find three pieces of information. Let us examine each question.

During which month did Maria's rocket club launch the most rockets?

Line graphs are good at showing changes in data over time. They are easy to use for making comparisons. On the graph above, the ticks at the bottom divide the graph into months. From the vertical axis to the first tick, the data is for May, the first month. The space between the second and third ticks is for June, etc. If we look at the data on the chart, the highest point takes place in the space between the two marks dividing off the month of July. So, July was the month with the most rocket launches!

What is the fewest number of rockets that the club launched during an active month?

It can sometimes be difficult to find exact numeric values on a line graph unless they are marked precisely. If you look at the first point on the chart, it is difficult to tell if that point shows 7 or 8 launches for the first day of May. If you were to take a ruler and measure the distance, you could tell that the point is below halfway between the 5 and the 10, so it shows 7 launches. But that is quite a bit of work! Luckily, the question we are asked is about the lowest point on the graph. The graph ends on the line for the number 5. Five is the fewest number of rockets that the club launched during an active month!

What trends do you notice in the data?

A **trend** is a pattern that can be seen in a set of data over time. Trends can help people make predictions about future data in the same set. Look at the graph at the top of this page. The trend in the early months was for the rocket launches to increase. In July, the launches peaked, and after that, there is a downward trend for the rest of the season. Can you think of a reason that fewer rockets might be launched at the beginning and ending of the season than during the middle?

73

Finding Missing Data on a Line Graph

There is a lot of data that can be shown on line graphs. Sometimes an exact measurement can be found, and other times you have to use estimation to figure out what the data says. This graph shows George's height from the time he was born until he was 10 years old. Using the graph, find how tall George was when he was $7\frac{1}{2}$ years old.

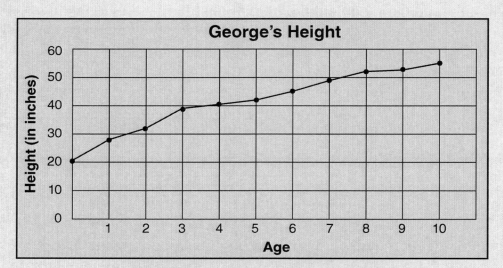

Step 1: Find the point on the horizontal axis that is midway between 7 and 8. Use a straight edge to measure directly up from that point to the place where it intersects with the graph.

Step 2: Using the straight edge again, find the point on the vertical axis directly to the left of this point.

When George was $7\frac{1}{2}$ years old, he was 50 inches tall.

Analyzing Line Graphs in Our Daily Lives

Because line graphs are good for showing change over time, they are often used in finance. Let us say a person owns part of a company. This is called "stock." At different times that stock is worth different amounts. If a person wants to make money with their stock, they should try to buy it when it is inexpensive and sell it when it is worth more. A line graph can help them track how their stock is doing. This can help them know when to buy more and when to sell.

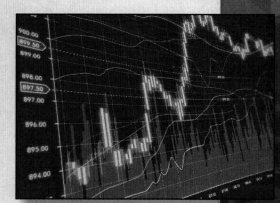

You Try It

Use the graph above to find how tall George was when he was $4\frac{1}{2}$ years old.

Analyzing Line Graphs

Maria's model rocket club convenes during sunny weather to launch the rockets that they construct. This graph shows how many rocket launches the group participated in during each month this year. During which month did Maria's club launch the most rockets? What is the fewest number of rockets that the club launched during an active month? What trends do you observe in this data?

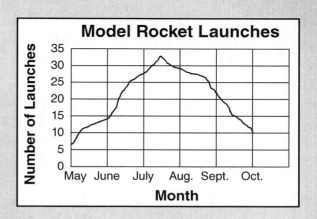

Basic Facts

The problem asks us to find three pieces of information. We will examine each question individually.

During which month did Maria's rocket club launch the most rockets?

Line graphs express longitudinal changes in data and are especially useful for making comparisons. On this graph, the ticks divide the horizontal axis into months. From the vertical axis to the first tick represents the month of May. Between the second and third ticks is June, etc. The highest point is established in the space between the two marks delineating the month of July; therefore, July was the month with the most rocket launches.

What is the fewest number of rockets that the club launched during an active month?

It can be difficult to determine exact numeric values on a line graph if they are not marked precisely. Examine the first point on the chart; it is unclear if it shows 7 or 8 launches for the first day of May. If you were to take a ruler and measure the distance, you could determine more precisely that the point is not quite halfway between the 5 and the 10, and conclude that it shows approximately 7 launches. That can be laborious! Luckily, this question asked about the lowest point on the graph. The graph concludes on the line for the number 5, so 5 is the fewest number of rockets that the club launched during an active month!

What trends do you notice in the data?

A **trend** is a pattern that can be observed in a set of data over time. Trends can help people make predictions about future data in the same set. Look at the graph at the top of this page. The trend in the early months was for the rocket launches to increase. In July, the launches peaked, and after that, there is a downward trend for the rest of the season. Can you think of a reason that fewer rockets might be launched at the beginning and ending of the season than during the middle?

Finding Missing Data on a Line Graph

Many different kinds of data can be displayed on line graphs. Sometimes an exact measurement can be found, and other times you have to use estimation to figure out what the data says. This graph shows George's height from the time he was born until he was 10 years old. Using the graph, determine how tall George was when he was $7\frac{1}{2}$ years old.

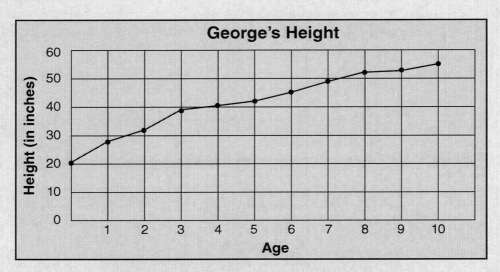

George's Height

Step 1: Find the point on the horizontal axis that is exactly midway between 7 and 8. Use a straight edge to measure directly up from that point to the place where it intersects with the graph.

Step 2: Using the straight edge again, find the point on the vertical axis directly to the left of this point.

When George was $7\frac{1}{2}$ years old, he was 50 inches tall.

Analyzing Line Graphs in Our Daily Lives

Line graphs are used in finance because they show changes over time. Imagine a person owns stock in a company. At different times that stock is worth different amounts. To make money with the stock, buy it when it is inexpensive and sells it when it is worth more. A line graph is a tool that can assist people in tracking how the stock is doing, when to buy more, and when to sell.

You Try It

Use the graph above to find how tall George was when he was $4\frac{1}{2}$ years old.

#50755—Leveled Texts for Mathematics: Data Analysis and Probability   © Shell Education

Creating Circle Graphs

Every student in Mr. James' class must pass a test. The test is on multiplication. They must pass it by the end of the year. He gave the test each quarter. He tracked the number of students who passed each time.

Multiplication Masters

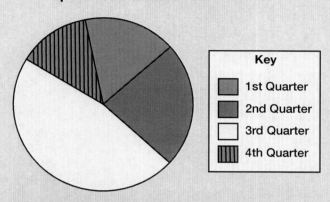

Key	
▨	1st Quarter
▨	2nd Quarter
☐	3rd Quarter
▥	4th Quarter

In what quarter did the most students pass the test?

Basic Facts

Circle graphs are named for their shapes. They are circles. They are used when the data shows parts of a whole. They are good for showing proportions. A **proportion** is a comparison of a part to a whole. You can look at the relationship between the sizes of things.

A percentage protractor is a special tool. It helps draw circle graphs. It shows the relationship between the degrees in a circle. It shows the percentage of the circle. For instance, $180°$ is marked as 50% because it is $\frac{1}{2}$ of a circle.

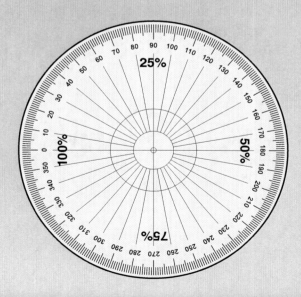

Creating a Circle Graph

Circle graphs can be based on percentages. They can also use fractions. They are easier to make if you have a percentage protractor. Look at the problems below. We will use numbers that are easy to work with. You will not need any special tools.

There are 120 members of the dance team. In the show, 60 members will wear red. There are 30 purple costumes. There are 15 black costumes. And 15 members will wear yellow. Make a circle graph to show all the colors.

Step 1: **Decide if the data is part of a whole**. If not, we should not make a circle graph! Here we know the whole is 120.

$60 + 30 + 15 + 15 = 120$. So, this data set is fine.

Step 2: **Convert your data to percentages**. To do this, divide each smaller section by the whole.

$60 \div 120 = 0.5 = 50\%$

$30 \div 120 = .25 = 25\%$

$15 \div 120 = .125 = 12.5\%$

(Don't worry, this is just $\frac{1}{2}$ of 25% so it will be easy to draw.)

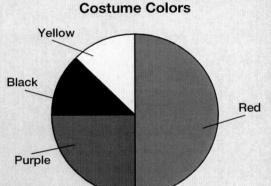

Costume Colors

Step 3: **Create the graph**. Draw a circle first. Then, make the sections. Try to be as precise as you can.

Step 4: **Label each section.** Give your graph a title.

Circle Graphs in Our Daily Lives

Circle graphs show data. They compare parts. They are used in politics. They show many things. They show numbers of votes. They show how many people like an idea. People study the graphs. Then they know how people will vote.

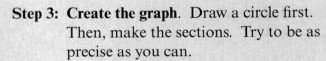

You Try It

Make a circle graph for this data.

Ja'Quan's math team is taking a vote. They are choosing a name for their team. There are 10 votes for "Mathemagicians." There are 15 votes for "The Math Team." There are 20 votes for "Humble Pi." And, there are five votes for "The Addin' Dragons."

Creating Circle Graphs

Every student in Mr. James' class must pass a test. The test is on multiplication facts. And, they must pass it by the end of the year. He gave the test each quarter. He kept track of the number of students who passed each time.

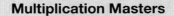

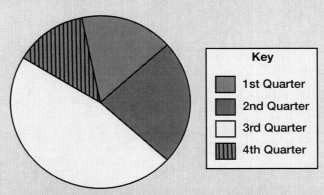

In what quarter did the most students pass the test?

Basic Facts

Circle graphs are named because of their shape. They are used when the data shows parts of a whole. They are good for showing proportions. A **proportion** is a comparison of a part to a whole. It helps you look at the relationship between the sizes of things.

A percentage protractor is a special tool. It helps draw circle graphs. It shows the relationship between the degrees in a circle and the percentage of the circle. For instance, 180° is marked as 50% because it is $\frac{1}{2}$ of a circle.

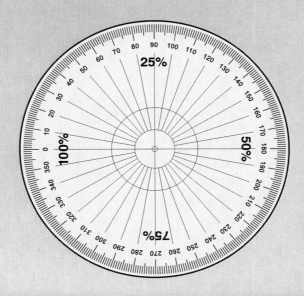

#50755—Leveled Texts for Mathematics: Data Analysis and Probability

Creating a Circle Graph

Circle graphs can be based on percentages. Or they can use fractions. And, they are easier to make if you have a percentage protractor. Look at the problems below. We will use numbers that should be easy to work with. So you won't need any special tools.

There are 120 members of the dance team. For their show, 60 of them will be wearing red costumes. There are 30 purple costumes. There are 15 black costumes. And 15 of them will be wearing yellow costumes. Make a circle graph to show all the colors.

Step 1: **Decide if the data is part of a whole**. If not, it will not be right for making a circle graph! In this case we know the whole is 120.

60 + 30 + 15 + 15 = 120. So, this data set is fine.

Step 2: **Convert your data to percentages**.
To do this, divide each subsection by the whole.
60 ÷ 120 = 0.5 = 50%
30 ÷ 120 = .25 = 25%
15 ÷ 120 = .125 = 12.5%

(Don't worry, this is just $\frac{1}{2}$ of 25% so it will be easy to draw.)

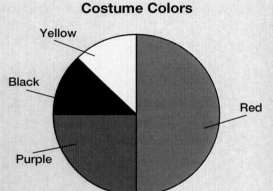
Costume Colors

Step 3: **Create the graph**. Draw a circle first. Then, make the sections. Try to be as accurate as you can.

Step 4: **Label the sections**. Give your graph a title.

Circle Graphs in Our Daily Lives

Circle graphs show how data compares to a whole. They are a good tool in politics. They are good at showing how many people like a candidate. And, they can show which people are in favor of an idea.

You Try It

Make a circle graph for this data.

Ja'Quan's math team is voting on their favorite team name. There are 10 people who vote for "Mathemagicians." There are 15 votes for "The Math Team." There are 20 votes for "Humble Pi." And, there are five votes for "The Addin' Dragons."

#50755—*Leveled Texts for Mathematics: Data Analysis and Probability*   © *Shell Education*

Creating Circle Graphs

Every student in Mr. James' class must pass a multiplication facts test before the end of the year. Mr. James gave the test each quarter and kept track of the number of students who passed each time.

Multiplication Masters

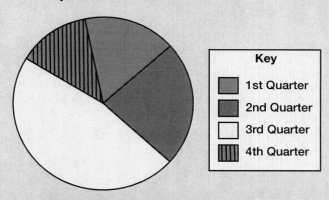

During which quarter did the most students pass the exam?

Basic Facts

Circle graphs are named because of their shape. They are used when the data being displayed represents parts of a whole. Circle graphs are useful for comparing proportions. A **proportion** is a comparison of a part to a whole. With proportions, you are looking at the relationship between the sizes of things.

A percentage protractor is a special tool for drawing circle graphs. It shows the relationship between the degrees in a circle and the percentage of the circle. For instance, 180° is marked as 50% because it is $\frac{1}{2}$ of a circle.

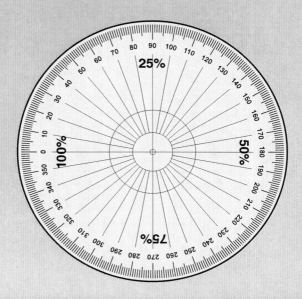

Creating a Circle Graph

Circle graphs can be based on percentages or fractions. Either way, they are easier to make if you have a percentage protractor. For the examples on this page, we will use numbers that should be easy to work with. You won't need any special tools.

There are 120 members of the dance team. For their upcoming performance, 60 of them will be wearing red costumes, 30 will be wearing purple costumes, 15 will be wearing black costumes, and 15 will be wearing yellow costumes. Make a circle graph to show the costume colors.

Step 1: Decide if the data is part of a whole. If not, it will not be suitable for making a circle graph! In this case we know the whole is 120.

60 + 30 + 15 + 15 = 120, so this data set is fine.

Step 2: Convert your data to percentages.
To do this, divide each subsection by the whole.

$60 \div 120 = 0.5 = 50\%$

$30 \div 120 = .25 = 25\%$

$15 \div 120 = .125 = 12.5\%$

(Don't worry, this is just $\frac{1}{2}$ of 25% so it will be easy to draw.)

Costume Colors

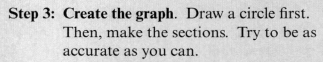

Step 3: Create the graph. Draw a circle first. Then, make the sections. Try to be as accurate as you can.

Step 4: Label the sections. Give your graph a title.

Circle Graphs in Our Daily Lives

Circle graphs show how data compares to a whole. Circle graphs are a popular tool in politics. They are good at showing what percentage of the population supports a candidate. Or they can show what portion of the population is in favor of a particular idea.

You Try It

Make a circle graph for the following data.

Ja'Quan's math team is voting on their favorite team name. There are 10 people who vote for "Mathemagicians." There are 15 people who vote for "The Math Team." There are 20 people who vote for "Humble Pi." There are five people who vote for "The Addin' Dragons."

#50755—*Leveled Texts for Mathematics: Data Analysis and Probability*   © *Shell Education*

Creating Circle Graphs

Every student in Mr. James' class is expected to pass a multiplication facts test prior to the end of the year. Mr. James gave a test each quarter and tracked the number of students who passed each time.

Multiplication Masters

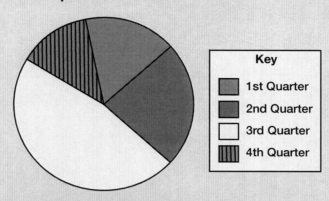

Key

- 1st Quarter
- 2nd Quarter
- 3rd Quarter
- 4th Quarter

During which quarter did the most students pass the exam?

Basic Facts

Circle graphs are circles, which explains where they get their names. They are used when the information being displayed represents parts of a whole. Circle graphs are extremely useful for comparing proportions. A **proportion** is a comparison of a part to a whole. With proportions, you are looking at the relationship between the sizes of things.

A percentage protractor is a special tool for drawing circle graphs. It shows the relationship between the degrees in a circle and the percentage of the circle. For instance, 180° is marked as 50% because it is $\frac{1}{2}$ of a circle.

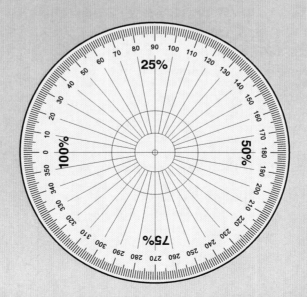

© *Shell Education* #50755—*Leveled Texts for Mathematics: Data Analysis and Probability*

Creating a Circle Graph

Circle graphs can be based on percentages or fractions, but either way, they are easier to make if you have a percentage protractor. However, for the examples on this page, we will use numbers that should be easy enough to work with and that don't require any special tools.

There are 120 members of the dance team. For their upcoming performance, 60 of them will be wearing red costumes, 30 will be wearing purple costumes, 15 will be wearing black costumes, and 15 will be wearing yellow costumes. Create a circle graph that displays the percentages of costume colors.

Step 1: **Decide if the data is part of a whole**. If not, it will not be suitable for making a circle graph! In this case we know the whole is 120.

60 + 30 + 15 + 15 = 120, so this data set is fine.

Step 2: **Convert your data to percentages**.
To do this, divide each subsection by the whole.

$60 \div 120 = 0.5 = 50\%$

$30 \div 120 = .25 = 25\%$

$15 \div 120 = .125 = 12.5\%$

(Don't worry, this is just $\frac{1}{2}$ of 25% so it will be easy to draw.)

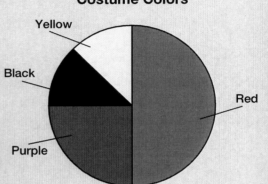

Costume Colors

Yellow · Black · Purple · Red

Step 3: **Create the graph**. Draw a circle first and then make the sections. Try to be as accurate as you can.

Step 4: **Label the sections**. Give your graph a title.

Circle Graphs in Our Daily Lives

Circle graphs show how data compares to a whole. Circle graphs are a popular tool in politics because they are excellent at displaying what percentage of the population supports a candidate. They can also illustrate what portion of the population is in favor of a particular idea.

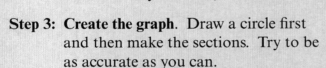

You Try It

Create a circle graph displaying the following data.

Ja'Quan's math team is voting on their favorite team name. There are 10 people who vote for "Mathemagicians." There are 15 people who vote for "The Math Team." There are 20 people who vote for "Humble Pi." There are five people who vote for "The Addin' Dragons."

84

Analyzing Circle Graphs

Hans made $200 this month. He did this by mowing lawns. This graph shows how he plans to spend his money. Each part has a percentage. On what does Hans plan to spend the least money? How does Hans plan to spend $\frac{1}{4}$ of his money? How much money does Hans plan to spend on video games?

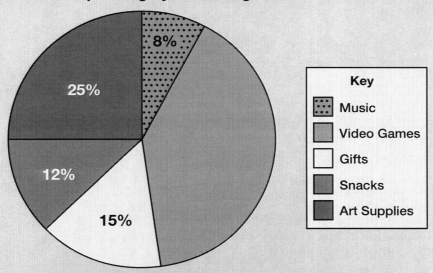

Hans's Spending by Percentage

Key
Music
Video Games
Gifts
Snacks
Art Supplies

Basic Facts

The problem asks us to find three pieces of data. Let us look at each question.

On which item does Hans plan to spend the least money?

Circle graphs show how data compares to a whole. In this case, the whole is all the money that Hans will spend. This graph has the sections labeled. But one label is missing. Look at the graph. It is easy to see that the part for music is the smallest. This means that Hans plans to spend the least on music.

How does Hans plan to spend $\frac{1}{4}$ of his money?

There is an easy way to find this answer. Look at the parts of the graph. There are three parts smaller than $\frac{1}{4}$. They are gifts, snacks, and music. The part for video games is larger than $\frac{1}{4}$. And the part for art supplies seems to take up $\frac{1}{4}$ of the circle. We can easily see this because the graph is labeled. We know that 25% is equal to $\frac{1}{4}$. So Hans plans to spend $\frac{1}{4}$ of his money on art supplies.

Basic Facts (cont.)

How much does Hans plan to spend on video games?

Now we have to find the missing part of the graph. The video game section is not labeled. But the other sections are.

Step 1: Use the information that you do have.

What do we know? We know that the whole graph shows all of his money. That is 100%.

We can subtract to find what percent is missing.

$$100 - (8 + 15 + 12 + 25) = 100 - (60) = 40$$

Step 2: Convert the number to the form asked for.

We are not done. We know that Hans will spend 40% of his money on video games. But we need to find an amount. So we have to take it a step further.

Hans made $200 total and we need to find 40% of that.

40% of $200 is equal to $0.4 \times 200 = \$80$

Hans plans to spend $80 on video games!

Analyzing Circle Graphs in Our Daily Lives

Circle graphs are good for showing parts of a whole. So they are often used to show how money is spent. Think of a youth club that sells candy to raise money. They want people to know that the money they make is for the club. Then people feel like their donation is helpful. The club might make a circle graph showing how their money is spent. This will help people see how their purchase will help the club.

You Try It

This graph shows the number of each kind of pet at the Placertown Pet Store. Look at the data on the graph. Can you estimate how many fish are being sold at the store?

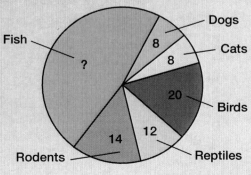

Number of Pets by Type

Fish — ?
Dogs — 8
Cats — 8
Birds — 20
Reptiles — 12
Rodents — 14

#50755 *Leveled Texts for Mathematics: Data Analysis and Probability*   © Shell Education

Analyzing Circle Graphs

Hans made $200 this month. He mowed lawns. The graph shows his plans to spend the money. Each part has a percentage. Where will Hans spend the least money? How does Hans plan to spend $\frac{1}{4}$ of his money? How much money does Hans plan to spend on video games?

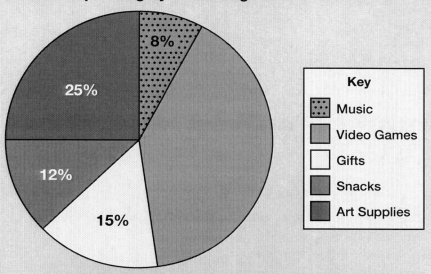

Hans's Spending by Percentage

Basic Facts

The problem asks us to find three pieces of data. Let us look at each question.

On which item does Hans plan to spend the least money?

Circle graphs show how data compares to a whole. In this case, the whole is all the money that Hans will spend. This graph has the sections labeled. But one label is missing. Look at the graph. It is easy to see that the part for music is the smallest. This means that Hans plans to spend the least on music.

How does Hans plan to spend $\frac{1}{4}$ of his money?

There is an easy way to find this answer. Look at the parts of the graph. The sections for gifts, snacks, and music are all smaller than $\frac{1}{4}$. The section for video games is larger than $\frac{1}{4}$. And, the section for art supplies seems to take up $\frac{1}{4}$ of the circle. We can easily see this because the graph is labeled. We know that 25% is equal to $\frac{1}{4}$. So Hans plans to spend $\frac{1}{4}$ of his money on art supplies.

Basic Facts *(cont.)*

How much does Hans plan to spend on video games?

Now we have to find the missing part of the graph. The video game section is not labeled. But the other sections are.

Step 1: Use the information that you do have.

We know that the entire graph represents 100% of Hans' money.

We can subtract to find what percent is missing.

$100 - (8 + 15 + 12 + 25) = 100 - (60) = 40$

Step 2: Convert the number to the form asked for.

We are not done. We know that Hans will spend 40% of his money on video games. But the question is asking for an amount. So we have to take it a step further.

Hans earned $200 all together and we need to find 40% of that.

40% of $200 is equal to $0.4 \times 200 = \$80$

Hans plans to spend $80 on video games!

Analyzing Circle Graphs in Our Daily Lives

Because circle graphs are good for showing parts of a whole, they are often used to illustrate how money is spent. For instance, think of a youth club that sells candy as a fundraiser. They want people to know that most of the money from the products they are selling goes directly to the club because they want people to feel like their donation is useful. The club might make a circle graph showing how their money is spent. This will help people see how their purchase will help the club.

You Try It

This graph shows the number of each kind of pet at the Placertown Pet Store. Given the data on the graph, about how many fish are being sold at the store?

Number of Pets by Type

- Fish — ?
- Dogs — 8
- Cats — 8
- Birds — 20
- Reptiles — 12
- Rodents — 14

Analyzing Circle Graphs

Hans earned $200 this month mowing lawns. This graph shows how he plans to spend his money by percentage. On which item does Hans plan to spend the least money? How does Hans plan to spend $\frac{1}{4}$ of his money? How much money does Hans plan to spend on video games?

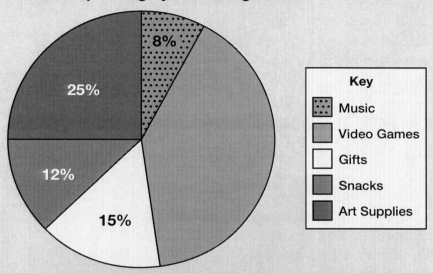

Hans's Spending by Percentage

Key
- Music
- Video Games
- Gifts
- Snacks
- Art Supplies

Basic Facts

The problem asks us to find three pieces of information. Let us examine each question.

On which item does Hans plan to spend the least money?

Circle graphs are good at showing how data compares to a whole. In this case, the whole is all the money that Hans will spend. This graph has the sections labeled, but one label is missing. In looking at the graph, though, it is easy to see that the section that represents music is the smallest. This means that Hans plans to spend the least on music.

How does Hans plan to spend $\frac{1}{4}$ of his money?

The easiest way to find the answer to this question is to look at the sections of the graph. The sections for gifts, snacks, and music are all smaller than $\frac{1}{4}$. The section for video games is larger than $\frac{1}{4}$ and the section for art supplies seems to take up $\frac{1}{4}$ of the circle. We can easily verify this because the graph is labeled by percentage. We know that 25% is equal to $\frac{1}{4}$, so Hans plans to spend $\frac{1}{4}$ of his money on art supplies.

© Shell Education   #50755—*Leveled Texts for Mathematics: Data Analysis and Probability*

Basic Facts *(cont.)*

How much does Hans plan to spend on video games?

This question asks us to find missing information. The video game section is not labeled, but the other sections are.

Step 1: Use the information that you do have.

We know that the entire graph represents 100% of Hans' money.

We can subtract to find what percent is missing.

$$100 - (8 + 15 + 12 + 25) = 100 - (60) = 40$$

Step 2: Convert the number to the form asked for.

We are not done. We know that Hans will spend 40% of his money on video games, but the question is asking for an amount so we have to take it a step further.

Hans earned $200 all together and we need to find 40% of that.

40% of $200 is equal to $0.4 \times 200 = \$80$

Hans plans to spend $80 on video games!

Analyzing Circle Graphs in Our Daily Lives

Because circle graphs are good for showing parts of a whole, they are often used to illustrate how money is spent. For instance, think of a youth club that sells candy as a fundraiser. They want people to know that most of the money from the products they are selling goes directly to the club because they want people to feel like their donation is useful. The club might make a circle graph showing how their money is spent so that people can easily see how their purchase will help the organization.

You Try It

This graph shows the number of each kind of pet at the Placertown Pet Store. Given the data on the graph, about how many fish are being sold at the store?

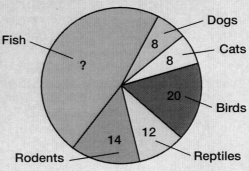

Number of Pets by Type

Analyzing Circle Graphs

Hans earned $200 last month mowing lawns and decides to spend his earnings. This circle graph displays how he plans to allocate his money by percentage. On which item does Hans plan to spend the least money? How does Hans plan to spend $\frac{1}{4}$ of his money? How much money does Hans estimate he'll spend on video games?

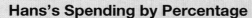

Hans's Spending by Percentage

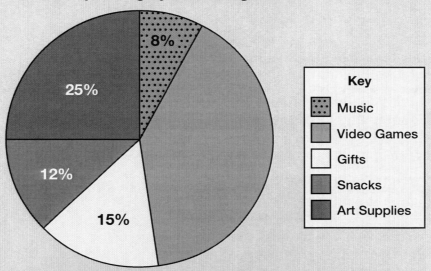

Basic Facts

The problem asks us to find three pieces of information. Let us examine each question separately.

On which item does Hans plan to spend the least money?

Circle graphs are excellent at portraying how individual pieces of data compare to a whole. In this example, the whole represents all the money that Hans will spend. This graph has the sections labeled with the percentages, but one label is missing. However, by just looking at the graph, it is easy to see that the section that represents music is the smallest, so that means that Hans plans to spend the least money on music.

How does Hans plan to spend $\frac{1}{4}$ of his money?

The easiest way to find the answer to this question is to look at the individual sections of the graph. The sections for gifts, snacks, and music are all smaller than $\frac{1}{4}$. The section for video games is larger than $\frac{1}{4}$ and the section for art supplies seems to take up $\frac{1}{4}$ of the circle. We can easily verify this because the graph is labeled by percentage. We know that 25% is equal to $\frac{1}{4}$, so Hans plans to spend $\frac{1}{4}$ of his money on art supplies.

© Shell Education   #50755—Leveled Texts for Mathematics: Data Analysis and Probability

Basic Facts *(cont.)*

How much does Hans plan to spend on video games?

This question asks us to find missing information. The video game section is not labeled but the other sections are, so we can start with those percentages.

Step 1: Use the information that you do have.

We know that the entire graph represents 100% of Hans' money.

We can subtract to find what percent is missing.

$100 - (8 + 15 + 12 + 25) = 100 - (60) = 40$

Step 2: Convert the number to the form asked for.

We have not answered the question yet. We know that Hans will spend 40% of his money on video games, but the question is asking for a monetary amount so we have to take it a step further.

Hans earned $200 all together and we need to find 40% of that.

40% of $200 is equal to $0.4 \times 200 = \$80$

Hans plans to spend $80 on video games!

Analyzing Circle Graphs in Our Daily Lives

Circle graphs are excellent at displaying parts of a whole, so they are frequently used to illustrate how money is spent. For example, think of a youth club that sells candy as a fundraiser. They want people to know that the majority of the money from the products they are selling goes directly to the club because they want people to feel like their donation is useful. The club might make a circle graph showing how their money is spent so that people can easily see how their purchase will help fund the organization.

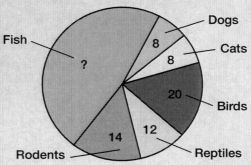

You Try It

This circle graph displays the number of each type of pet sold at the Placertown Pet Store. Given the information displayed on the graph, about how many fish are being sold at the store?

Number of Pets by Type

Fish: ?
Dogs: 8
Cats: 8
Birds: 20
Reptiles: 12
Rodents: 14

#50755—*Leveled Texts for Mathematics: Data Analysis and Probability*   © Shell Education

Comparing Graphs

Look at the graphs. Which one shows how many students in Mr. Tannith's class got an "A" on the science test?

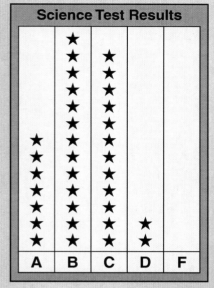

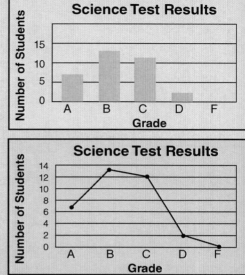

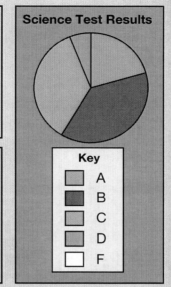

Basic Facts

Look at the graphs at the top of the page. Each one shows the same set of data. But not all data works well with each kind of graph. The circle graph shows the same data. But it is not as clear as the bar graph and pictograph. Each graph type has strengths. And each one has weaknesses.

Pictograph

Pictographs are good at comparing sets of data. They are best used when the data can be shown as multiples of the same number. They are also good for small data sets. You can have each picture be one data point.

Bar Graph

Bar graphs are good at comparing sets of data. They can also be used to track large changes over time. But they are not good at showing small change over time.

Line Graph

Line graphs are good at showing change over time. They are the best choice when data is always changing. They are good to use when the changes are small.

Circle Graph

Circle graphs are good at showing parts of a whole.

© Shell Education   #50755—Leveled Texts for Mathematics: Data Analysis and Probability

Choosing a Graph

Look at a few sets of data. We will choose which graph would be best to show it.

Example 1: Temperature readings for the month of March

This data set shows change over time. So a line graph would be a good choice.

Example 2: The number of books checked out by students from each class in the school

This data set shows a comparison. A bar graph might be a good choice. You could also use a circle graph. This would show what percentages of books each class checked out.

Example 3: There are 14 students wearing blue shirts. There are 7 red shirts. There are 21 black shirts. And there are 7 white shirts.

All the data in the example are multiples of 7. A pictograph might be a good choice here.

Example 4: There are 12 dogs at a shelter. There are six puppies below the age of 2. There are four between the ages of 2 and 6. And there are two over the age of 6.

This data may be best shown as parts of a whole. A circle graph would be a good choice. A bar graph or pictograph might also be good choices.

Graph Choices in Our Daily Lives

Have you ever seen a graph in a newspaper? Or have you played a computer game that showed your scores on a graph? Have you been watching the news and seen a graph? Where else have you seen graphs? People make graphs all the time. They have data that they want to present. So they choose the best graph for their data.

You Try It

Wilma is in an art class. Her class took a field trip to the zoo. Look at the chart. It shows how many students drew each kind of animal. Choose which graph best shows the data. Then draw it.

Tiger	Zebra	Elephant	Tarantula	Gorilla
10	12	5	2	7

#50755—*Leveled Texts for Mathematics: Data Analysis and Probability*   © Shell Education

Comparing Graphs

Look at the graphs. Which one shows how many students in Mr. Tannith's class got an "A" on the science test?

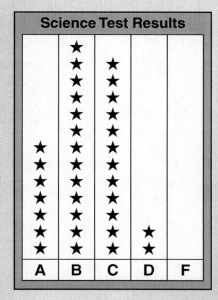

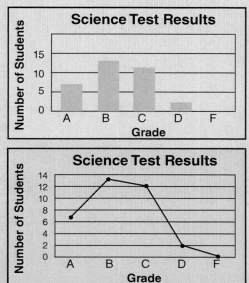

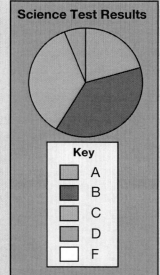

Basic Facts

Look at the graphs at the top of the page. Each one shows the same set of data. But not every set of data is presented well with each type of graph. The circle graph above shows the same data. But it is not as clear as the bar graph and pictograph. Each graph type has strengths. And each one has weaknesses.

Pictograph

Pictographs are good for showing comparisons between sets of data. They are best used when the data can be shown as multiples of the same number. They are also good for small data sets where each picture can be one data point.

Bar Graph

Bar graphs are good for showing comparisons between sets of data. They can also be used to track large changes over time. But because of their scale, they are not good at tracking smaller changes over time.

Line Graph

Line graphs are good for showing changes in data over time. They are the best choice when data is constantly changing.   They are good to use when the changes are small.

Circle Graph

Circle graphs are good for showing data sets that are parts of a whole.

© Shell Education   #50755—Leveled Texts for Mathematics: Data Analysis and Probability

Choosing a Graph

Let us look at several sets of data. Then we will choose which graph type would be best to show it.

Example 1: Temperature readings for the month of March

This data set shows change over time. So a line graph would be a good choice.

Example 2: The number of library books checked out by students from each class in the school

This data set shows a comparison. A bar graph might be a good choice. Or if you want to show what percentage of the total number of books each class checked out, a circle graph could be used.

Example 3: There are 14 students wearing blue shirts. There are 7 red shirts. There are 21 black shirts. And there are 7 white shirts.

All the data in the example are multiples of 7. A pictograph might be a good choice here.

Example 4: There are 12 dogs at a shelter. There are six puppies below the age of 2. There are four between the ages of 2 and 6. And there are two over the age of 6.

This data may be best shown as a comparison of the parts to the whole. A circle graph would be a good choice. A bar graph or pictograph might also be good choices.

Graph Choices in Our Daily Lives

Have you seen graphs in a newspaper? Or played a computer game that showed your scores on a graph? Have you seen graphs on the news? Where else have you seen graphs? For every graph someone had data to present. They thought that a graph would be a good way to show it. Each of those people had to choose what graph would be best for their data.

You Try It

Wilma is in an art class. This chart shows how many students chose to draw each type of animal during their field trip to the zoo. Decide which graph best shows the data and draw it.

Tiger	Zebra	Elephant	Tarantula	Gorilla
10	12	5	2	7

#50755—*Leveled Texts for Mathematics: Data Analysis and Probability*

Comparing Graphs

Which graph shows how many students in Mr. Tannith's class earned an "A" on the science test?

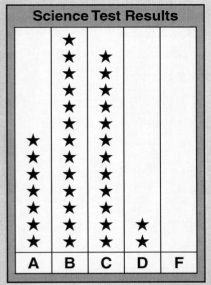

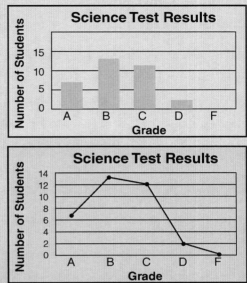

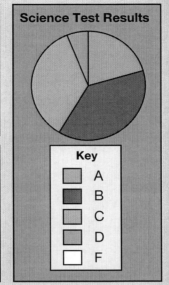

Basic Facts

Look at the examples at the top of the page. Each graph displays identical sets of data. But not every type of graph is the best presentation for that set of data. The circle graph above presents the same data, for instance. But it is not as clear as the bar graph and pictograph. Each graph type has strengths and limitations.

Pictograph

Pictographs are good for showing comparisons between sets of data. They are most suitable when the data can be presented as multiples of the same number. They are also useful for small data sets where each picture can be one data point.

Bar Graph

Bar graphs are good for showing comparisons between sets of data. They can also be used to track large changes over time. But because of their scale, they are not usually good at tracking smaller changes over time.

Line Graph

Line graphs are good for showing changes in data over time. They are usually the best choice when data is constantly changing or when the changes are small.

Circle Graph

Circle graphs are good for showing data sets that are parts of a whole.

97

Choosing a Graph

Let us look at several sets of data and decide which graph type would be best to show it.

Example 1: Temperature readings for the month of March

This data set shows change over time, so a line graph would be a good choice.

Example 2: The number of library books checked out by students from each class in the school

This data set shows a comparison. A bar graph might be a good choice. Or if you want to emphasize what percentage of the total number of books each class checked out, a circle graph could be used.

Example 3: There are 14 students wearing blue shirts, 7 wearing red shirts, 21 wearing black shirts, and 7 wearing white shirts.

All the data in the example are multiples of 7. A pictograph might be a good choice here.

Example 4: There are 12 dogs at a shelter. There are six puppies below the age of 2, four between the ages of 2 and 6, and two over the age of 6.

This data may be best displayed as a comparison of the parts to the whole. A circle graph would be a good choice. A bar graph or pictograph might also be good choices.

Graph Choices in Our Daily Lives

Have you ever seen a graph in a newspaper or played a computer game that showed you your scores on a graph? Have you been watching the news and seen a graph? Where else have you seen graphs? Someone made every graph that you see. Someone had data to present and then decided that a graph would be a good way to show it. Each of those people had to decide what graph would be best for their data.

You Try It

Wilma is in an art class. This chart shows how many students chose to draw each type of animal during their field trip to the zoo. Decide which graph best shows the data and draw it.

Tiger	Zebra	Elephant	Tarantula	Gorilla
10	12	5	2	7

#50755—*Leveled Texts for Mathematics: Data Analysis and Probability*　　　© *Shell Education*

Comparing Graphs

Which of the graphs below shows how many students in Mr. Tannith's class got an "A" on the science test?

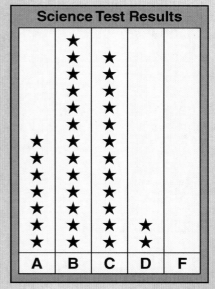

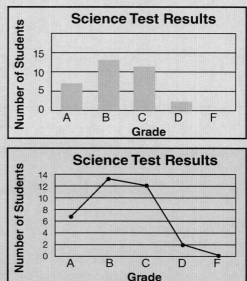

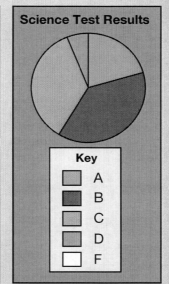

Basic Facts

Look at the examples above. Each graph displays the same set of data but not every set of data is presented well with each type of graph. For example, the circle graph above presents the same information but it is not as clear or easy to comprehend as the bar graph and pictograph. Each different graph type has strengths and limitations.

Pictograph

Pictographs are excellent for showing comparisons between sets of data. They are most suitable when the data can be presented as multiples of the same number. They are also useful for small data sets where each picture can be one data point.

Bar Graph

Bar graphs are good for showing comparisons between sets of data and can also be used to track large changes over time. But because of their scale, they are not usually good at tracking smaller changes over time.

Line Graph

Line graphs are excellent for showing changes in data over time. They are usually the best choice when data is constantly changing or when the changes are small.

Circle Graph

Circle graphs are good for showing data sets that are parts of a whole.

#50755—Leveled Texts for Mathematics: Data Analysis and Probability

Choosing a Graph

Let us look at several sets of data and decide which graph type would be the most effective choice.

Example 1: Temperature readings for the month of March

This data set shows change over time, so a line graph would be a good choice.

Example 2: The number of library books checked out by students from each class in the school

This data set shows a comparison, so a bar graph might be a good choice. Or if you want to emphasize what percentage of the total number of books each class checked out, a circle graph could be used.

Example 3: There are 14 students wearing blue shirts, 7 wearing red shirts, 21 wearing black shirts, and 7 wearing white shirts.

All the data in the example are multiples of 7, so a pictograph might be a good choice here.

Example 4: There are 12 dogs at a shelter. There are six puppies below the age of 2, four between the ages of 2 and 6, and two over the age of 6.

This data may be best displayed as a comparison of the parts to the whole, so a circle graph would be a good choice. A bar graph or pictograph might also be good choices.

Graph Choices in Our Daily Lives

Have you ever seen a graph in a newspaper or played a computer game that displayed your scores on a graph? Have you been watching the news and seen a graph? Where else have you seen graphs? Someone created every graph that you see. Someone had information to present and then decided that a graph would be the perfect way to illustrate it. Each of those people had to decide what type of graph would be best for their data.

You Try It

Wilma is in an art class. This chart shows how many students chose to draw each type of animal during their field trip to the zoo. Decide which graph best illustrates the data and draw it.

Tiger	Zebra	Elephant	Tarantula	Gorilla
10	12	5	2	7

100

What Does Mean Mean?

What did the bar graph say to the line graph?

Basic Facts

Joe spends an average of 1 hour a day on his homework. Ella is about average height. The average snowfall is 7 feet per year.

What is average? When something is "average" it stands for a group. **Average** is a way to use one number to get an idea about a group of data. In math you are asked to find the average. This means to find the mean. To find the **mean** of a group of numbers, add the numbers. Then divide the sum by the number of addends. The mean is one way to use a number to stand for the whole group.

Strengths and Weaknesses of the Average

Let's look at some instances of average. This will help us understand its strengths and weaknesses.

Example 1: Bernard sells skis in his shop. He still needs to sell an average of five pairs of skis per day. This will be enough money to pay his rent. This chart shows Bernard's sales for the week. Did he average enough sales to pay rent?

Monday	Tuesday	Wednesday	Thursday	Friday	Saturday	Sunday
7	4	5	4	6	7	9

$7 + 4 + 5 + 4 + 6 + 7 + 9 = 42$ $42 \div 7 = 6$

Yes! Bernard did sell enough skis this week.

Look at the numbers in the list. Does the mean seem to represent the group? Why or why not? Is there another way to show this group with one number?

Example 2: Shelly likes to skip rope. She keeps track of how long she skips rope each day. Here is a chart of her results for five days..

Day 1	Day 2	Day 3	Day 4	Day 5
25 min.	27 min.	3 min.	24 min.	26 min.

$25 + 27 + 3 + 24 + 26 = 105$ $105 \div 5 = 21$

Shelly averaged 21 minutes per day.

101

Strengths and Weaknesses of the Average (cont.)

Look at the data from the previous page. One number was much smaller than the rest. This is an **outlier**. It means there is a big gap between one number and the rest. Look at the mean. Look at the list of numbers in the set. The mean was smaller than most of the other numbers. The outlier is not like the other data.

Do you think the mean still represents the group? Can we find a way to make sure that a single number does not change the results?

Finding the Mean

There are two steps to find the mean. Find the mean for this set of numbers: 300, 250, 375, 140, 218, 337.

> **Step 1:** Find the sum of all the numbers in the group.
>
> $300 + 250 + 375 + 140 + 218 + 337 = 1,620$
>
> **Step 2:** Divide the sum by the number of addends.
>
> $1,620 \div 6 = 270$
>
> The mean of this set of data is 270.

Using Mean in Our Daily Lives

There are a lot of stats kept on baseball players. Batting average is not used the same as most means. It is shown as a decimal. It is rounded to the thousandths. Imagine a batter has been at bat 20 times. He hit the ball 10 times. His batting average would be $10 \div 20 = .500$. We say his batting average is 500.

You Try It

Jonas has been keeping track of his points. He did this for every basketball game this season. Here is a chart of his results. Use this data to find the mean of Jonas' scores.

Game 1	Game 2	Game 3	Game 4	Game 5	Game 6
24	14	30	23	19	28

(on average, people are mean!)

What Does Mean Mean?

What did the bar graph say to the line graph?

Basic Facts

Joe spends an average of 1 hour a day on his homework. Ella is about average height. The average snowfall for an area is seven feet per year.

What is average? When something is "average" it stands for a group. **Average** is a way to use a single item to get an idea about a group of data. In math you are asked to find the average. This means you are being asked to find the mean. The **mean** of a group of numbers is the sum of the group divided by the number of addends. Finding the mean is one way to find a number that stands for the whole group.

Strengths and Weaknesses of the Average

Look at some examples of averages. These will help us understand the strengths and weaknesses.

Example 1: Bernard sells skis in his shop. He needs to sell an average of five pairs of skis per day. This will be enough money to pay his rent. This chart shows Bernard's sales for the week. Did he average enough sales to pay rent?

Monday	Tuesday	Wednesday	Thursday	Friday	Saturday	Sunday
7	4	5	4	6	7	9

$7 + 4 + 5 + 4 + 6 + 7 + 9 = 42$ $42 \div 7 = 6$

Yes! Bernard did sell enough skis this week.

Look at the numbers in the list. Does the mean seem to represent the group? Why or why not? Can you think of another way to represent this group with one number?

Example 2: Shelly is keeping track of how long she skips rope each day. Here is a chart of her results for five days.

Day 1	Day 2	Day 3	Day 4	Day 5
25 min.	27 min.	3 min.	24 min.	26 min.

$25 + 27 + 3 + 24 + 26 = 105$ $105 \div 5 = 21$

Shelly averaged 21 minutes per day.

        #50755—Leveled Texts for Mathematics: Data Analysis and Probability

Strengths and Weaknesses of the Average *(cont.)*

In this case, one of the numbers in the group was much smaller than the rest. This is called an **outlier**. That is because there is a big gap between this number and the rest. Look at the mean. Now look at the list of numbers in the set. The mean turned out to be smaller than most of the numbers in the set. This is because the outlier is not like the other data.

Do you think the mean still represents the group? How could we make sure that a single number is less likely to skew the results?

Finding the Mean

Computing the mean of any group of numbers is a two-step process. Let us find the mean for this set of numbers: 300, 250, 375, 140, 218, 337.

Step 1: Find the sum of all the numbers in the group.

300 + 250 + 375 +140 + 218 + 337 = 1,620

Step 2: Divide the sum by the number of addends.

1,620 ÷ 6 = 270

The mean of this set of data is 270.

Using Mean in Our Daily Lives

There are a lot of stats kept on baseball players. One is their batting average. Batting average is not used the same as most means. It is written as a decimal. It is rounded to the thousandths. Think of a batter who had been at bat 20 times and hit the ball 10 times. His batting average would be 10 ÷ 20 = .500. We say his batting average is 500.

You Try It

Jonas has been keeping track of his points. He did this for every basketball game this season. Here is a chart of his results. Use this data to find the mean of Jonas' scores.

Game 1	Game 2	Game 3	Game 4	Game 5	Game 6
24	14	30	23	19	28

What Does Mean Mean?

What did the bar graph say to the line graph?

Basic Facts

Joe spends an average of one hour each day on his homework. Ella is about average height. The average snowfall for an area is seven feet per year.

What is average? When something is "average" it represents a group. **Average** is a way to use a single item to get a general idea about an entire group of data. In mathematics if you are asked to find the average, you are usually being asked to find the mean. The **mean** of a group of numbers is the sum of the group divided by the number of addends. Finding the mean is one way to determine a number that represents the entire group.

Strengths and Weaknesses of the Average

Let's look at some examples of average to understand its strengths and weaknesses.

Example 1: Bernard sells skis in his shop. He needs to sell an average of five pairs of skis per day to earn enough money for his rent. This chart shows Bernard's sales for the week. Did he average enough sales to pay rent?

Monday	Tuesday	Wednesday	Thursday	Friday	Saturday	Sunday
7	4	5	4	6	7	9

$$7 + 4 + 5 + 4 + 6 + 7 + 9 = 42 \qquad\qquad 42 \div 7 = 6$$

Yes! Bernard did sell enough skis this week.

Look at the numbers in the list. Does the mean seem to represent this group? Why or why not? Can you think of another way to represent this group with one number?

Example 2: Shelly is recording how long she skips rope each day. Here is a chart of her results after tracking for five days.

Day 1	Day 2	Day 3	Day 4	Day 5
25 min.	27 min.	3 min.	24 min.	26 min.

$$25 + 27 + 3 + 24 + 26 = 105 \qquad\qquad 105 \div 5 = 21$$

Shelly averaged 21 minutes per day.

 #50755—Leveled Texts for Mathematics: Data Analysis and Probability

Strengths and Weaknesses of the Average *(cont.)*

In this case, one of the numbers in the group was much smaller than the rest. This is an **outlier**. That is because there is a large gap between this number and the rest. Return to the mean and compare it to the list of numbers in the set. In actuality, the mean was smaller than most of the numbers in the set. This is because the outlier was much different than the other data.

Do you think the mean still represents the whole group? How could we be assured that a single number won't skew the overall results?

Finding the Mean

Computing the mean of any group of numbers is a two-step process. Let us practice by finding the mean for this set of numbers: 300, 250, 375, 140, 218, 337.

Step 1: Find the sum of all the numbers in the group.

$$300 + 250 + 375 + 140 + 218 + 337 = 1,620$$

Step 2: Divide the sum by the number of addends.

$$1,620 \div 6 = 270$$

The mean of this set of data is 270.

Using Mean in Our Daily Lives

One of the statistics that is tracked about baseball players is their batting averages. This number is calculated by taking their total number of hits divided by their total number of times at bat. Batting average is not used the same as most means. It is written as a decimal and then rounded to the thousandths. In other words, a batter who had been at bat 20 times and hit the ball 10 times would use $10 \div 20 = .500$. We say his batting average is 500.

You Try It

Jonas has been keeping track of his points in every basketball game this season. Here is a chart of his results. Use this data to find the mean of Jonas' scores.

Game 1	Game 2	Game 3	Game 4	Game 5	Game 6
24	14	30	23	19	28

(on average, people are mean!)

What Does Mean Mean?

What did the bar graph say to the line graph?

Basic Facts

Joe spends an average of 1 hour a day on his homework. Ella is about average height. The average snowfall for a certain region is typically seven feet per year.

What is average? **Average** is a way for a single item to establish a general idea about the entire group of data. If you are asked to find the average, you are usually being asked to determine the mean. To find the **mean** of a group of numbers means to find the sum of the numbers and then divide by the number of addends. Finding the mean of a group of numbers is one method for finding a number that represents the entire group.

Strengths and Weaknesses of the Average

Let's investigate examples of average to understand its strengths and limitations.

Example 1: Bernard sells skis in his shop and he needs to sell an average of five pairs of skis per day to earn enough money for his rent. The following chart displays Bernard's sales for the week. Did Bernard average enough sales in order to pay rent?

Monday	Tuesday	Wednesday	Thursday	Friday	Saturday	Sunday
7	4	5	4	6	7	9

$7 + 4 + 5 + 4 + 6 + 7 + 9 = 42$ $42 \div 7 = 6$

Yes! Bernard did sell enough skis this week.

Analyze the numbers in the list. In this example, does the mean seem to accurately represent the group? Why? Determine another way to represent this group with one number.

Example 2: Shelly is keeping track of how long she skips rope each day. Here is a chart of her results for five days.

Day 1	Day 2	Day 3	Day 4	Day 5
25 min.	27 min.	3 min.	24 min.	26 min.

$25 + 27 + 3 + 24 + 26 = 105$ $105 \div 5 = 21$

Shelly averaged 21 minutes per day.

Strengths and Weaknesses of the Average *(cont.)*

In this example, one of the numbers in the group was significantly smaller than the rest. This is an **outlier** because there is a substantial gap between this number and the rest. Look again at the mean and then compare it to the list of numbers in the set. The mean turned out to be considerably smaller than most of the numbers in the set because there is such a distance between the outlier and the other numbers.

Do you still think the mean accurately represents the group? How could we be assured that one single number would not skew the results to affect the outcome?

Finding the Mean

Computing the mean or average of any group of numbers is a two-step procedure. Let us practice by determining the mean for the following set of numbers: 300, 250, 375, 140, 218, 337.

Step 1: Find the sum of all the numbers in the group.

$$300 + 250 + 375 + 140 + 218 + 337 = 1,620$$

Step 2: Divide the sum by the number of addends.

$$1,620 \div 6 = 270$$

The mean of this set of data is 270.

Using Mean in Our Daily Lives

One of the statistics that is monitored for baseball players is their batting average. Batting average is calculated by taking the total number of a player's hits divided by the total number of times at bat. Batting average is not determined in the same way as most means. First, it is written as a decimal so it can be rounded to the nearest thousandth. So, if there was a batter who had been at bat 20 times and had made contact with the ball 10 times, his batting average would be $10 \div 20 = .500$. We say his batting average is 500.

You Try It

Jonas has been keeping track of his points in every basketball game this season. Below is a chart that displays his results. Use this information to determine the mean of Jonas' scores.

Game 1	Game 2	Game 3	Game 4	Game 5	Game 6
24	14	30	23	19	28

(on average, people are mean!)

Median in the Middle

A **median** is something that is in the middle. Some roads have a median strip. This is a piece of land that keeps the cars going one way apart from the cars going the other way.

Basic Facts

Let's say you have a group of numbers. The median is the number in the middle. It then stands for the group of numbers. There are times where it is more useful than the mean. Think about this case.

Mira tests video games for a living. Her team tries to find mistakes, or "bugs," in the code. They do this every day. They find the bugs. And then they fix them. Here is a chart of how many bugs Mira's team found each day.

Day 1	Day 2	Day 3	Day 4	Day 5	Day 6	Day 7
17	21	15	117	19	20	22

On the 4th day they had tried to fix something. But instead, they had made some mistakes. That caused a lot of bugs! Mira's team found many more bugs that day. There were 117 bugs that day! This is called an outlier in the data. There is a big gap between it and the rest of the numbers in the group.

First let us find the mean.

$$17 + 21 + 15 + 117 + 19 + 20 + 22 = 231$$

$$231 \div 7 = 33$$

Does 33 seem to represent the group? Could she say that her team finds about 33 bugs per day? Most days her team finds less than 33 bugs. But the 4th day skews the results. That number is not at all like the rest. So it throws off the mean. Now let us find the median. List the numbers in order. The median is the middle number. So let's list the numbers of bugs from least to greatest. It will look like this:

$$15, 17, 19, \mathbf{20}, 21, 22, 117$$

There are 7 numbers. So the middle number is the 4th one. In this case, it is the number 20. Does the number 20 seem to represent the group? Could she say that her team finds about 20 bugs each day? In this case the median shows the data better than the mean.

Finding the Median for an Even Set of Numbers

Let's say you have an odd number of numbers. Then finding the median is simple. Put the group in numeric order. Then find the number in the middle. But what if there are an even amount of numbers? In that case there are two numbers in the middle. Here are the steps to follow to find the median in an even set of numbers: 102, 114, 125, 100, 99, 116.

Step 1: List the data in numeric order.

99, 100, 102, 114, 116, 125

Step 2: Find the two middle numbers.

99, 100, **102, 114**, 116, 125

Step 3: Then find the number that is in between. You need to find the mean of the two middle numbers.

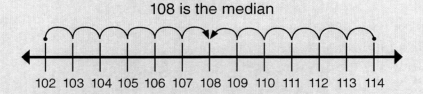

108 is the median

102 + 114 = 216

216 ÷ 2 = 108

108 is the median of the set.

Median in Our Daily Lives

Medians are good for looking at data about groups. Think about people who live in a city. Some of them may be rich. But the rest of the people might not have much money. Because of the outliers, the mean income may be quite high. It will look like the people are all well off. But this is not a good representation of the whole city. So the median would be the better choice.

You Try It

Find the median for this set of numbers: 14, 35, 19, 72, 24, 26, 11, 28

Median in the Middle

A **median** is something that is in the middle. Some roads have a median strip. This is a piece of land that keeps the cars going one way apart from the cars going the other way.

Basic Facts

Median is another way for one number to stand for a group of numbers. The median is the middle value of a set of numbers. There are times where it is more useful than the mean. Think about this case.

Mira tests video games for a living. Her team tries to find mistakes, or "bugs," in the code. They do this every day. They find the bugs. And then they fix them. Here is a chart of how many bugs Mira's team found each day.

Day 1	Day 2	Day 3	Day 4	Day 5	Day 6	Day 7
17	21	15	117	19	20	22

On the 4th day they had tried to fix something. But instead, they had made some mistakes. That caused a lot of bugs! Mira's team found many more bugs than usual. The 4th day's result of 117 bugs is called an outlier in the data. There is a big gap between it and the rest of the numbers in the group.

First let us find the mean.

$$17 + 21 + 15 + 117 + 19 + 20 + 22 = 231$$

$$231 \div 7 = 33$$

Does 33 represent the group? Did Mira's team find about 33 bugs per day? Most days they find less than 33. But the 4th day changes the results. That number throws off the average. Now find the median. It is the middle number when the set is listed in numerical order. List the number of bugs found from least to greatest, like this:

15, 17, 19, **20**, 21, 22, 117

There are 7 numbers. So, the middle number is the 4th in the series. In this case, it is the number 20. Does the number 20 seem to represent the group? Could Mira say that her team finds about 20 bugs each day? In this case the median shows the data better than the mean.

Finding the Median for an Even Set of Numbers

Finding the median for an odd number of numbers is simple. Put the group in numeric order. Then find the number in the middle. But what if there are an even amount of numbers? In that case there are two numbers in the middle. Here are the steps to find the median in an even set of numbers: 102, 114, 125, 100, 99, 116.

Step 1: List the data in numeric order.

99, 100, 102, 114, 116, 125

Step 2: Find the two middle numbers.

99, 100, **102**, **114**, 116, 125

Step 3: Find the number halfway between the two middle numbers. Another way to think of this is that you are finding the mean of the two middle numbers.

108 is the median

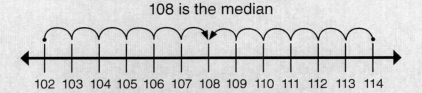

102 103 104 105 106 107 108 109 110 111 112 113 114

$102 + 114 = 216$

$216 \div 2 = 108$

108 is the median of the set.

Median in Our Daily Lives

Medians are good for looking at data about groups. Think about the incomes of people living in a city. There may be some rich people living in one part. But the rest of the city is full of people who do not make much money. Because of the outliers, the mean income may be quite high. It looks like the people are all well off. But this is not a good representation of the whole city. So the median would be the better choice.

You Try It

Find the median for this set of numbers: 14, 35, 19, 72, 24, 26, 11, 28

Median in the Middle

A **median** is something that is in the middle. On a highway or divided road, the median strip is a piece of land that keeps cars going one direction separated from cars going the other way.

Basic Facts

Median is another way for one number to represent a group of numbers. The median is the middle value of a set of numbers. Sometimes it can be a better representative than the mean, or average. Think about this example.

Mira tests video games for a living. Every day her team tries to find mistakes, or "bugs," in the code. When they find them, the engineers can fix them. Here is a chart of how many bugs Mira's team found each day.

Day 1	Day 2	Day 3	Day 4	Day 5	Day 6	Day 7
17	21	15	117	19	20	22

On the 4th day the engineers had tried to fix something. But instead, they had made some mistakes. That caused a lot of bugs! Mira's team found many more bugs than usual. The 4th day's result of 117 bugs is called an outlier in the data. There is a large gap between it and the rest of the numbers in the group.

First let us find the mean.

$$17 + 21 + 15 + 117 + 19 + 20 + 22 = 231$$

$$231 \div 7 = 33$$

Does 33 seem to represent the group? Could Mira say that her team typically finds about 33 bugs per day? Most days Mira's team finds far fewer than 33 bugs. But the 4th day skews the results. That means that because the one result is much different than the rest, it distorts the average.

Now let us find the median. The median is the middle number when the set is listed in numerical order. If we list the number of bugs found from least to greatest, it looks like this:

15, 17, 19, **20**, 21, 22, 117

There are 7 numbers. So the middle number is the 4th in the series. In this case, it is the number 20. Does the number 20 seem to represent the group? Could Mira say that her team finds about 20 bugs on a typical day? In this case the median is a better representative of the data than the mean.

113

Finding the Median for an Even Set of Numbers

Finding the median for an odd number of numbers is simple. Put the group in numeric order. Then find the number in the middle. But what if there are an even amount of numbers? In that case there are two numbers in the middle. Here are the steps for finding the median in an even set of numbers: 102, 114, 125, 100, 99, 116.

Step 1: List the data in numeric order.

99, 100, 102, 114, 116, 125

Step 2: Find the two middle numbers.

99, 100, **102**, **114**, 116, 125

Step 3: Find the number halfway between the two middle numbers. Another way to think of this is that you are finding the mean of the two middle numbers.

108 is the median

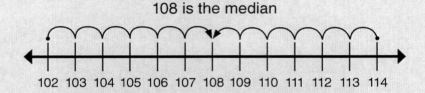

$$102 + 114 = 216$$

$$216 \div 2 = 108$$

108 is the median of the set.

Median in Our Daily Lives

Median values are often useful for understanding data about groups of people. Pretend you are looking at the income of people living in a city. There may be some very wealthy people living in one area. But the rest of the city is full of people who do not make very much money. Because of the few extreme outliers, the mean income may be quite high. It looks like the people are all well off. But this is not a very good representation of how most people in the city live. In this case, the median value would give you a better idea about most of the people living there.

You Try It

Find the median for this set of numbers: 14, 35, 19, 72, 24, 26, 11, 28

Median in the Middle

A **median** is something that is in the middle. On a highway or divided road, the median strip is a section of land that keeps cars going one direction separated from cars going the other way.

Basic Facts

Median is another way for one number to represent a group of numbers. The median is the middle value of a set of numbers. Sometimes it can be a better representative than the mean, or average. Let's investigate the following example.

Mira tests video games for a living. Every day her team tries to find mistakes, or "bugs," in the code. When they find these mistakes, the engineers can fix them. Here is a chart of how many bugs Mira's team found each day.

Day 1	Day 2	Day 3	Day 4	Day 5	Day 6	Day 7
17	21	15	117	19	20	22

On the 4th day the engineers had tried to fix something, but instead they made some mistakes. That caused a lot of bugs! Mira's team discovered many more bugs than usual. The 4th day's result of 117 bugs is called an outlier in the data. There is a substantial gap between it and the rest of the numbers in the group.

First let us find the mean.

$$17 + 21 + 15 + 117 + 19 + 20 + 22 = 231$$

$$231 \div 7 = 33$$

Does 33 seem to be an accurate number to represent the group? Could Mira say that her team typically uncovers about 33 bugs per day? Most days Mira's team finds far fewer than 33 bugs, but the 4th day skews the results. That means that because the one result is much different than the rest, it distorts the average.

Now let us find the median. The median is the middle number when the set is listed in numerical order. If we list the number of bugs found from least to greatest, it looks like this:

15, 17, 19, **20**, 21, 22, 117

There are 7 numbers, so the middle number is the 4th in the series. In this example, it is the number 20. Does the number 20 seem to be an accurate representation of the group? Could Mira say that her team finds about 20 bugs on a typical day? In this example the median is a better representative of the data than the mean.

115

Finding the Median for an Even Set of Numbers

Finding the median for an odd number of numbers is simple. You just arrange the group of numbers in numeric order and then find the number in the middle. However, what if there are an even amount of numbers? In that case there are two numbers in the middle. Here are the steps to follow to find the median in an even set of numbers: 102, 114, 125, 100, 99, 116.

Step 1: List the data in numeric order.

99, 100, 102, 114, 116, 125

Step 2: Find the two middle numbers.

99, 100, **102**, **114**, 116, 125

Step 3: Find the number halfway between the two middle numbers. Another way to think of this is that you are finding the mean of the two middle numbers.

108 is the median

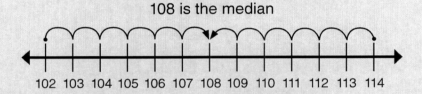

$102 + 114 = 216$

$216 \div 2 = 108$

108 is the median of the set.

Median in Our Daily Lives

Median values are often very useful for understanding data about groups of people. Pretend you are looking at the income of people living in a city. There may be some very wealthy people living in one area, but the remainder of the city is full of people who do not make very much money. Because of the few extreme outliers, the mean income may be quite high. It may appear as though the entire population of the city is well off. However, this is not a very accurate representation of how most people in the city live. In this example, the median value would give you a better idea about most of the people living there.

You Try It

Find the median for this set of numbers: 14, 35, 19, 72, 24, 26, 11, 28

Mode and Range

Have you ever had pie á la mode? *Á la mode* is a French term. It means "in style or fashion." Ice cream was in style when the term was first being used. So now *pie á la mode* means "pie with ice cream"! Do you know the song "Home on the Range"? A range is a big open space. In math, mode and range have different meanings. The meanings can help you remember the math.

Basic Facts

Range

There is **range** in math, too. It is the distance between the least and greatest numbers in a set. Look at this set of data: 1, 5, 7, 10. The range is 9. You subtract 1 from 10. To remember range, imagine a big open space between two end numbers.

Mode

The **mode** is the number that shows up the most often in a set. Some sets have more than one mode. Some sets have no mode. In the set {1, 4, 4, 3, 5, 5, 7} there are two modes. The numbers 4 and 5 are both modes of this set. They each show up two times. That is more than any other number. Remember, the mode means in fashion. That means many people are doing it. The mode appears more often than the others.

117

Finding the Mode and Range for a Data Set

We can find the mode. We can find the range. Monique and her friends make a record. They track how many servings of vegetables they eat in one week. Monique ate 15 servings. John ate 17 servings. Jonas ate 11 servings. Nico ate 7 servings. Patrick ate 22 servings. Sasha ate 17 and Geraldo ate 12. Thanh ate 23 and Kim ate 20. Laron ate 17 servings.

Step 1: List the data in numerical order.

7, 11, 12, 15, 17, 17, 17, 20, 22, 23

Step 2: Subtract the smallest number from the largest number.

23 − 7 = 16

16 is the range for this set of data.

Step 3: Find the mode. How many times does each number occur.

In this case, each number occurs only once, except 17. It occurs three times.

17 is the mode of this set of data.

Mode in Our Daily Lives

Remember mode means about the same as "in fashion." Mode can be used to talk about fashion sales. A clothing store wants to know which styles are most popular. First they record all the items that they sold. They look for the items that sold the most. This helps them see what is in style.

You Try It

Nori's wrestling teammates were weighed. Here are their weights (in pounds):

110, 99, 109, 125, 110, 137, 90, 103, 114, 117, 114, 134, 114, 110

Find the range and mode or modes for this set of data.

Mode and Range

Have you ever ordered pie á la mode? *Á la mode* is a French term. It means "in the current style or fashion." Ice cream was in style when the term was first being used. With time *pie á la mode* came to mean "pie with ice cream"! Have you ever sung the song "Home on the Range"? A range is a big open space. In math, mode and range have different meanings. Maybe you will see some things are the same as you learn more.

Basic Facts

Range

In math, the **range** can be thought of as the distance between the least and greatest members of a set. Look at this set of data: 1, 5, 7, 10. The range is 9. That is the difference between 1 and 10. It's easy to remember range. Just think of a big open space between two end numbers.

Mode

The **mode** is the number that shows up the most often in a set of data. Some sets have more than one mode. Some sets have no mode. In the set {1, 4, 4, 3, 5, 5, 7} there are two modes. The numbers 4 and 5 are both modes of this set. That is because they each show up two times. That is the most that any number shows up in this data set. One way to think about mode is to think that being the mode is in fashion. It's like having ice cream with your pie! If something is in fashion, more people are doing it. So if a number is the mode, there are more of that number than the others.

Finding the Mode and Range for a Data Set

Let's look at this case to learn how to find the mode and range for a set of data. Monique and her friends want to record how many servings of vegetables they ate over a school week. Here are their results: Monique ate 15 servings. John ate 17 servings. Jonas ate 11 servings. Nico ate 7 servings. Patrick ate 22 servings. Sasha ate 17 and Geraldo ate 12. Thanh ate 23 and Kim ate 20. Laron ate 17 servings.

Step 1: List the data in numerical order.

7, 11, 12, 15, 17, 17, 17, 20, 22, 23

Step 2: Find the range by subtracting the smallest number from the largest number.

$23 - 7 = 16$

16 is the range for this set of data.

Step 3: Find the mode. Count how many times each number occurs.

In this case, each number occurs only once, except 17. It occurs three times.

17 is the mode of this set of data.

Mode in Our Daily Lives

We said that mode means about the same as "in fashion." In real life, mode is often used to talk about fashion and sales. Let us say a clothing company wants to find out which styles are most popular. First they would record all of the items that they sold. Then, they would find the item or items that sold most often. This would give them a good idea of what items people like at the moment!

You Try It

Nori's wrestling team weighed in before the match. Here are their weights in pounds:

110, 99, 109, 125, 110, 137, 90, 103,
114, 117, 114, 134, 114, 110

Find the range and mode or modes for this set of data.

Mode and Range

Have you ever ordered pie á la mode? *Á la mode* is a French term that means "in the current style or fashion." Ice cream was in style when the term was first introduced, so *pie á la mode* came to mean "pie with ice cream"! Have you ever recited the song "Home on the Range"? A range is a big open space. In mathematics, mode and range have different meanings. Perhaps you'll see connections between these terms as you read.

Basic Facts

Range

In mathematics, the **range** can be thought of as the distance between the least and greatest members of a set. Look at this set of data: 1, 5, 7, 10. The range is 9 because that is the difference between 1 and 10. One way to remember range is to imagine it as the big open space between the two end numbers.

Mode

The **mode** is the number that occurs most often in a set of data. Some sets have more than one mode, while others have no mode. In the set {1, 4, 4, 3, 5, 5, 7} there are two modes. The numbers 4 and 5 are both modes of this set because they each occur two times, and twice is the most that any number recurs in this data set. One way to remember mode is to think that being the mode is fashionable—like having ice cream with your pie! If something is in fashion, more people are doing it. So if a number is the mode, it is showing up more often than the others.

Finding the Mode and Range for a Data Set

We can experiment with finding the mode and the range for a set of data by using the following example. Monique and her friends decided to record how many servings of vegetables they ate over a school week. Their results are listed here: Monique ate 15 servings; John ate 17; Jonas ate 11; Nico ate 7; Patrick ate 22; Sasha ate 17; Geraldo ate 12; Thanh ate 23; Kim ate 20; and Laron ate 17.

Step 1: List the data in numerical order.

7, 11, 12, 15, 17, 17, 17, 20, 22, 23

Step 2: Find the range by subtracting the smallest number from the largest number.

23 – 7 = 16

16 is the range for this set of data.

Step 3: Find the mode by determining how many times each number occurs.

In this case, each number occurs only once, except 17, which occurs three times.

17 is the mode of this set of data.

Mode in Our Daily Lives

In the example on the previous page, we said that mode means about the same as "in fashion." In real life, mode is often used in the fashion industry to determine sales. Imagine a clothing manufacturer wants to find out which of the current styles are most popular. First they would record all of the merchandise that they sold. Then they would find the individual item or items that sold most frequently. This would give them a good idea of which pieces of merchandise people are gravitating towards at the moment!

You Try It

Nori's wrestling team weighed in before the match. Here are their weights in pounds:

110, 99, 109, 125, 110, 137, 90, 103,
114, 117, 114, 134, 114, 110

Find the range and mode or modes for this set of data.

Mode and Range

Have you ever ordered pie á la mode? *Á la mode* is a French term that means "in the current style or fashion." Ice cream was popular when the term was first being used, so now *pie á la mode* means "pie with ice cream"! Have you ever sung the song, "Home on the Range"? A range is an immense open region. In mathematics, mode and range have different meanings. However, maybe you will see some similarities as you read.

Basic Facts

Range

In mathematics, the **range** can be thought of as the distance between the least and greatest extremes of a set. Look at this set of data: 1, 5, 7, 10. The range is 9 it represents the difference between 1 and 10. Remember this meaning of range by thinking of it as the big open space between the two end numbers.

Mode

The **mode** is the most frequently occurring number in a set of data. Some sets have more than one mode, while others have no mode. In the set {1, 4, 4, 3, 5, 5, 7} there are two modes. The numbers 4 and 5 are both modes of this set because they each occur two times, and twice is the most often that any number recurs in this data set. Remember mode by thinking of a trend that is most fashionable—like having ice cream with your pie! If something is in fashion, more people are following that trend; so if a number is the mode, it is trending more frequently than the others.

123

Finding the Mode and Range for a Data Set

Let's demonstrate how to determine the mode and the range for a set of data. Monique and her friends intended to analyze the quantities of vegetables they consumed during a school week. Their results are represented in this list: Monique consumed 15 servings; John consumed 17; Jonas consumed 11; Nico consumed 7; Patrick consumed 22; Sasha consumed 17; Geraldo consumed 12; Thanh consumed 23; Kim consumed 20; and Laron consumed 17.

Step 1: Order the data numerically.

7, 11, 12, 15, 17, 17, 17, 20, 22, 23

Step 2: Calculate the range by subtracting the smallest number from the largest number.

23 – 7 = 16

16 is the range for this set of data.

Step 3: Find the mode by counting how many times each number occurs.

In this case, each number occurs only once, except 17, which occurs three times.

17 is the mode of this set of data.

Mode in Our Daily Lives

In the previous example, we compared the mathematical meaning of mode to something being "in fashion." Coincidentally, mode is utilized to understand trends relating to sales. Imagine a clothing manufacturer who wants to discover the most popular styles. First they would record each piece of merchandise that they sold and then they would distinguish the specific items with the highest sales rates; this data would provide clear evidence as to the most popular pieces of merchandise at the moment.

You Try It

Nori's wrestling team had a mandatory weigh-in prior to the competition. The competitors' weight distributions in pounds are shown here:

110, 99, 109, 125, 110, 137, 90, 103,
114, 117, 114, 134, 114, 110

Determine the range and mode or modes for this set of data.

Probability of Events

Juan heard three weather reports today. The first one said, "There is a 25% chance of rain." The second one said, "Rain is unlikely today." The third one said, "There is a one in f our chance of rain today." Should Juan carry an umbrella today?

Basic Facts

We need to know how likely an event is to happen. This is the **probability** of an event. It means chance. You may know terms like *likelihood*. You may know *percentage*. To show chance we can use a ratio. We can use a fraction. Look at the weather reports above. All three showed the same chance of rain. In all three, rain is unlikely. Juan can leave his umbrella at home. There is a small chance of rain. If Juan wants to be sure, he should bring his umbrella.

Understanding Likelihood

Certain

How likely is an event to happen? Some events are **certain**. They will happen for sure. Roll a die. It is certain to land on either 1, 2, 3, 4, 5, or 6. These are the only numbers on the die!

Impossible

Then there are times when events are **impossible**. They cannot ever happen. If you roll a die, it cannot land on a zero. There is no zero on the die!

Likely and Unlikely

Some events are **likely**. That means that there is a high chance they will occur. Five out of the six sides of a die have numbers greater than 1 on them. Roll a die. It is likely that it will land on a number greater than 1. Some events have a low chance of happening. They are called **unlikely**. It is unlikely that your die will land on 6 every time.

Equally Likely

Some events have the same chance of occurring or not. They are **equally likely** to occur. Roll a die. There is a chance it will land on an odd number. There is the same chance it will land on an even number. There are three odd numbers on the die. There are three even numbers on the die. The chances are equally likely.

125

How Likely Is It?

We can find how likely an event is to occur. Look at a spinner.

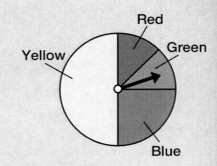

Spin the arrow. How likely is it to land on the green section? It is unlikely that the spinner will land on green. Green takes up only a small part of the circle.

How likely is the arrow to land on yellow? Yellow takes up $\frac{1}{2}$ of the space on the spinner. So it is equally likely that the spinner will land on yellow or that it will not.

How likely is the arrow to land on yellow or blue? The yellow + blue takes up the most space. The spinner is likely to land on one of those two colors.

How likely is it that the spinner will land on pink? Pink is not on the spinner! That is an impossible event.

How likely is it that the spinner will land on yellow, blue, red, or green? Those four colors cover the spinner. It is certain that the spinner will land on one of them.

Probability in Our Daily Lives

Every day, people make choices. They decide what to wear. They decide when to leave the house. They try to make good choices. Should you wear the long-sleeved shirt? Should you wear short sleeves? You need to know if it will be warm today. Should you be early to school? How likely is it that your friends will be there to play with? We choose based on the outcome we think will happen. That helps us make better choices.

You Try It

Taj is playing a word game. He pulls letter tiles. He cannot see them when he pulls them. Look at the choices below. Tell if each is *certain*, *likely*, *equally likely*, *unlikely*, or *impossible*.

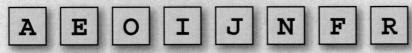

- Taj will pull the letter "*r*" on his next pick.
- Taj will pull the letter "*k*."
- Taj will pull a vowel.
- Taj will pull a letter from his name.

#50755—*Leveled Texts for Mathematics: Data Analysis and Probability*

Probability of Events

Juan heard three weather reports this morning. The first one said, "We have about a 25% chance of rain." The second one said, "Rain is unlikely today." The third one said, "There is a one in four chance that there will be rain today." Should Juan bring an umbrella to school?

Basic Facts

The **probability** of an event describes how likely it is to happen. It means chance. There are many ways to talk about this. It can be described with terms like *likelihood* or a *percentage*. It can be shown as a ratio. Or as a fraction. Look at the weather reports above. All three weather people are talking about the same chance of rain. In all three cases, rain is unlikely to occur. But it is not impossible. Juan can leave his umbrella at home. But he might get wet! If he wants to stay dry, he needs to bring his umbrella.

Understanding Likelihood

Certain

How likely is it that something will happen? Some events are **certain** to happen. That means they will occur for sure. If you roll a regular die, it is certain to land on a 1, 2, 3, 4, 5, or 6. That is because those are the only numbers on the cube!

Impossible

Then there are times when events are **impossible**. They cannot happen at all. If you roll a regular die, it cannot land on a zero. That is because there is no zero on the cube!

Likely and Unlikely

Some events are **likely**. This means that there is a high chance they will occur. Five out of the six sides of a regular die have numbers greater than 1 on them. So, if you roll a regular die, it is likely that it will land on a number greater than 1. Events with a low chance of happening are called **unlikely**. It is unlikely that your roll will land on number 6.

Equally Likely

Some events have the same chance of occurring as other events. Such sets of events are called **equally likely** to occur. If you roll a regular die there is an equally likely chance that it will land on an odd number as an even one. This is because there are three odd and three even numbers on the cube.

How Likely Is It?

We need to understand how likely a given event is to occur. Let us take a look at a spinner.

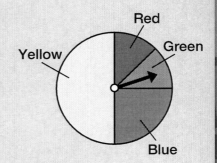

If you spin the arrow on this spinner, how likely is it that you will land on the green section? We can say that it is unlikely that the spinner will land on green. That is because green takes up only a small part of the space on the circle.

How likely is it that the spinner will land on yellow? Yellow takes up $\frac{1}{2}$ of the space on the spinner. So it is equally likely that the spinner will land on yellow or that it will not.

How likely is it that the spinner will land on either yellow or blue? The combination of yellow + blue takes up the most space on the spinner. So it is likely that the spinner will land on one of those two colors.

How likely is it that the spinner will land on pink? Pink is not on the spinner! So that is an impossible event.

How likely is it that the spinner will land on yellow, blue, red, or green? Those four colors cover the entire spinner. So it is certain that the spinner will land on one of them.

Probability in Our Daily Lives

Every day, people make choices. They decide what clothes to wear. They decide when to leave for work or school. They try to make choices that are good for them. Should you wear the long-sleeved shirt or the short-sleeved shirt? Well, how likely is it that it will be warm out today? Should you leave for school a little early? How likely is it that your friends will also be there early so you can talk with them? Each choice we make is based on what we think the outcome will be. Understanding probability can help us to make better choices.

You Try It

Taj is pulling letter tiles for a word game. The tiles are face down so he cannot see them. For each question, tell whether it is *certain*, *likely*, *equally likely*, *unlikely*, or *impossible*.

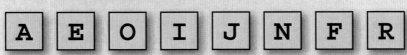

- Taj will pull the letter "*r*" on his next pick.
- Taj will pull the letter "*k*."

- Taj will pull a vowel.
- Taj will pull a letter from his name.

#50755—Leveled Texts for Mathematics: Data Analysis and Probability   © Shell Education

Probability of Events

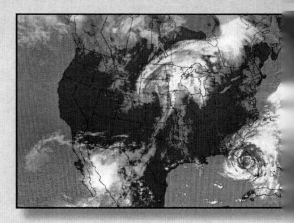

Juan heard three weather forecasts this morning. The first one said, "We have about a 25% chance of rain." The second reported, "Rain is unlikely today." The third weather person indicated that, "There is a probability of one in four that there will be rain today." Should Juan bring an umbrella to school?

Basic Facts

The **probability** of an event describes how likely it is to happen. There are many ways to talk about probability. It can be described with terms like *likelihood* or *percentage*. We can use a ratio or a fraction. In the example above, all three weather people are talking about the same chance of rain. In all three cases, rain is unlikely to occur, but it is not impossible. If Juan does not mind getting wet, he could choose to leave his umbrella at home. But if he wants to make sure that he stays dry, he should bring his umbrella—just in case!

Understanding Likelihood

Certain

How likely is it that something will happen? Some events are **certain** to happen. That means they will definitely occur. If you roll a regular die, it is certain to land on a 1, 2, 3, 4, 5, or 6 because those are the only numbers on the cube!

Impossible

Others are **impossible** events. They cannot happen at all. If you roll a regular die, it cannot land on a zero because there is no zero on the cube!

Likely and Unlikely

Some events are **likely**, which means that there is a high chance they will occur. Five out of the six sides of a regular die have numbers greater than 1 on them. So, if you roll a regular die, it is likely that it will land on a number greater than 1. Events with a low chance of happening are called **unlikely**. It is unlikely that your roll will land on number 6.

Equally Likely

Some events have the same chance of occurring as other events. Such sets of events are **equally likely** to occur. If you roll a regular die there is an equally likely chance that it will land on an odd number as an even one because there are three odd and three even numbers on the cube.

© Shell Education   #50755—Leveled Texts for Mathematics: Data Analysis and Probability

How Likely Is It?

It is important to understand how likely a given event is to occur. Let us take a look at a spinner.

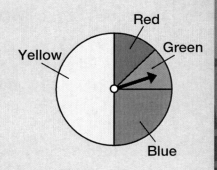

If you spin the arrow on this spinner, how likely is it that you will land on the green section? We can say that it is unlikely that the spinner will land on green, because green takes up only a small portion of the space on the circle.

How likely is it that the spinner will land on yellow? Yellow takes up $\frac{1}{2}$ of the space on the spinner. So it is equally likely that the spinner will land on yellow or that it will not.

How likely is it that the spinner will land on either yellow or blue? The combination of yellow + blue takes up a majority of the space on the spinner. So it is likely that the spinner will land on one of those two colors.

How likely is it that the spinner will land on pink? Pink is not on the spinner! So that is an impossible event.

How likely is it that the spinner will land on yellow, blue, red, or green? Those four colors cover the entire spinner. So it is certain that the spinner will land on one of them.

Probability in Our Daily Lives

Every day, people make decisions. They decide what clothes to wear. They decide when to leave for work or school. They decide each of these things based on the likelihood that their choices will benefit them. Should you wear the long-sleeved shirt or the short-sleeved shirt? Well, how likely is it that it will be warm out today? Should you leave for school a little early? How likely is it that your friends will also be there early so you can talk with them? Each choice we make is based on our best predictions about what the outcome will be. Understanding probability can help us to make better choices.

You Try It

Taj is pulling letter tiles for a word game. The tiles are face down so he cannot see them. For each scenario, tell whether it is *certain*, *likely*, *equally likely*, *unlikely*, or *impossible*.

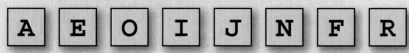

- Taj will pull the letter "*r*" on his next pick.
- Taj will pull a vowel.
- Taj will pull the letter "*k*."
- Taj will pull a letter from his name.

#50755—*Leveled Texts for Mathematics: Data Analysis and Probability* © *Shell Education*

Probability of Events

Juan heard three weather forecasts this morning. The first reporter said, "We have about a 25% chance of rain." The second forecast reported, "Rain is unlikely today." The third weather person indicated that, "There is a probability of one in four that there will be rain today." Should Juan bring an umbrella to school based on these predictions?

Basic Facts

The **probability** of an event describes how likely that event is to happen. Probability can be discussed in many different ways. Probability can be described in terms like *likelihood* or a *percentage*. It can be written as a ratio or as a fraction. In the example above, all three weather people are reporting about the same chance of rain. In all three examples, rain is unlikely to occur, but it is not impossible. If Juan does not mind getting wet, he may decide to leave his umbrella at home. But if he wants to make sure that he stays dry, he should definitely bring his umbrella—just in case!

Understanding Likelihood

Certain

How likely is it that something will happen? Some events are **certain** to happen, which means they will definitely occur. If you roll a regular die, it is absolutely certain to land on a 1, 2, 3, 4, 5, or 6 because those are the only numbers on the cube!

Impossible

Others are **impossible** events because they simply cannot happen at all. If you roll a regular die, it cannot land on a zero because there is no zero on the cube!

Likely and Unlikely

Some events are **likely**, which means that there is a high probability that they will occur. Five out of the six sides of a regular die have numbers greater than 1 on them. So if you roll a regular die, it is likely that it will land on a number greater than 1. Events with a low probability of happening are called **unlikely**. It is unlikely that your roll will land on number 6.

Equally Likely

Some events have the same probability of occurring as other events. Such sets of events are **equally likely** to occur. If you roll a regular die there is an equally likely chance that it will land on an odd number as an even number because there are three odd and three even numbers on the cube.

How Likely Is It?

It is important to understand how likely a given event is to occur. For example, let us take a look at a spinner.

If you spin the arrow on this spinner, how likely is it that you will land on the green section? We can say that it is unlikely that the spinner will land on green, because green takes up only a small portion of the space on the circle.

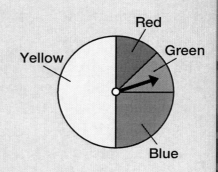

How likely is it that the spinner will land on yellow? Yellow takes up $\frac{1}{2}$ of the space on the spinner, so it is equally likely that the spinner will land on yellow or that it will not land on yellow.

How likely is it that the spinner will land on either yellow or blue? The combination of yellow + blue takes up a majority of the space on the spinner, so it is likely that the spinner will land on one of those two colors.

How likely is it that the spinner will land on pink? Pink is not even on the spinner, so landing on pink is an impossible event!

How likely is it that the spinner will land on yellow, blue, red, or green? Those four colors cover the entire spinner, so it is absolutely certain that the spinner will land on one of those colors.

Probability in Our Daily Lives

Every day, people make decisions. They decide what clothes to wear and they decide when to leave for work or school. People decide each of these things based on the likelihood that their choices will benefit them. Should you wear the long-sleeved shirt or the short-sleeved shirt? What's the probability that it will be warm outside today? Should you leave for school a little early this morning? How likely is it that your friends will also be there early so you can talk with them? Every choice we make is based on our best predictions about what the outcome will be. Therefore, understanding probability can help us to make better choices.

You Try It

Taj is pulling letter tiles for a word game. The tiles are face down so he cannot see them. For each scenario, tell whether the probability is *certain*, *likely*, *equally likely*, *unlikely*, or *impossible*.

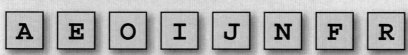

- Taj will pull the letter "*r*" on his next pick.
- Taj will pull the letter "*k*."
- Taj will pull a vowel.
- Taj will pull a letter from his name.

Probability Experiments

Lin and Jon are playing a game. There are four marbles inside a jar. One marble is purple. Three are yellow. They close their eyes. They pull a marble out of the jar. What is the probability that they will pull out the purple marble?

Basic Facts

Probability shows how likely an event is to occur. But what if we want a more exact answer? It can be shown as a fraction. Or it can be shown as a percentage. Say we want to find the probability. To do this, we look at the favorable outcomes. These are the results we want. Then we compare them with the total number of possible outcomes.

Look at the problem at the top of the page. In this case, they want to pull a purple marble out of the jar. There is only one case where that will happen. So there is one favorable outcome. Now count the total number of outcomes. There are four marbles total. So there are four possible outcomes.

The probability of an event can be shown in words.

There is a one-in-four chance that they will pull out a purple marble.

The probability of an event can be shown as a fraction in simplest form.

Favorable outcomes = 1

Possible outcomes = 4

There is a $\frac{1}{4}$ chance that they will pull out a purple marble.

The probability of an event can be shown as a percentage.

$1 \div 4 = .25 = 25\%$

There is a 25% chance that they will pull out a purple marble.

Finding the Probability

Use this spinner for the problems below. Find the probability that the spinner will land on blue.

Step 1: Find the possible number of favorable outcomes for your problem.

Count the blue spaces. We find that there are 5.

Step 2: Find the number of possible outcomes.

Count the total spaces. We find that there are 10.

Step 3: Find the fractional probability. Express it in simplest form.

$$\frac{5}{10} = \frac{1}{2}$$

There is a probability of $\frac{1}{2}$ that the spinner will land on blue.

Step 4: Find the percentage probability.

$$5 \div 10 = 0.5 = 50\%$$

There is a 50% probability that the spinner will land on blue.

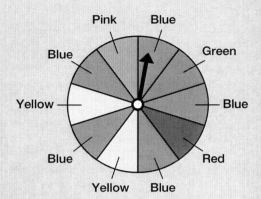

Probability Experiments in Our Daily Lives

Have you ever played a card game? How about a game with spinners or dice? Have you used a die with 20 sides? Some games use dice with 100 sides! These games are all based on probability. People who make the games know the chance for every outcome. That is how they try to make the game fun and fair. People who create video games do this, too. But in video games, the computer takes care of the probability. You may never see it happen!

You Try It

Look at the spinner above. Use it to answer the questions.

What is the probability that the spinner will land on yellow? Show your answer as a fraction.

What is the probability that the spinner will land on pink? Show your answer as a percentage.

134

Probability Experiments

Lin and Jon are playing a game. There are four marbles inside a jar. One marble is purple. Three are yellow. They pull a marble out of the jar without looking. What is the probability that they will pull out the purple marble?

Basic Facts

Probability shows how likely an event is to occur. But what if we want a more exact answer? In many cases it can be shown as a fraction. Or it can be shown as a percentage. Say we want to find the probability. To do this, we look at the number of favorable outcomes. These are the results we want. Then we compare them with the total number of possible outcomes.

Look at the problem at the top of the page. In this case, the favorable outcome is to pull a purple marble out of the jar. There is only one case where that will happen. There is one favorable outcome. Now count the total number of outcomes. There are four marbles total. So there are four possible outcomes.

The probability of an event can be expressed in words.

There is a one-in-four chance that Lin and Jon will pull out a purple marble.

The probability of an event can be expressed as a fraction in simplest form.

Favorable outcomes = 1   Possible outcomes = 4

There is a $\frac{1}{4}$ chance that Lin and Jon will pull out a purple marble.

The probability of an event can be expressed as a percentage.

$1 \div 4 = .25 = 25\%$

There is a 25% chance that Lin and Jon will pull out a purple marble.

© *Shell Education*   *#50755—Leveled Texts for Mathematics: Data Analysis and Probability*

Finding the Probability

Use this spinner for the problems below. Find the probability that the spinner will land on blue.

Step 1: Find the possible number of favorable outcomes for the condition of your problem.

Count the blue spaces. We find that there are 5.

Step 2: Find the number of possible outcomes.

Count the total spaces. We find that there are 10.

Step 3: Find the fractional probability. Express it in simplest form.

$$\frac{5}{10} = \frac{1}{2}$$

There is a probability of $\frac{1}{2}$ that the spinner will land on blue.

Step 4: Find the percentage probability.

$$5 \div 10 = 0.5 = 50\%$$

There is a 50% probability that the spinner will land on blue.

Probability Experiments in Our Daily Lives

Have you ever played a card game? How about a game with spinners or dice? Maybe you have played a game that uses special dice that have 20 sides. Some games even use dice with 100 sides! These games are all based on probability. The game designers know the exact chance for every possible outcome. They use this knowledge to try to make the game fun and fair for everyone. People who create video games do this too. But in video games, the computer takes care of the probability. The players may never see it happen!

You Try It

Look at the spinner above. Use it to answer the questions.

What is the probability that the spinner will land on yellow? Show your answer as a fraction.

What is the probability that the spinner will land on pink? Show your answer as a percentage.

Probability Experiments

Lin and Jon are playing a game. There are four marbles inside a jar. One marble is purple and three are yellow. They pull a marble out without looking. What is the probability that they will pull out the purple marble?

Basic Facts

Probability shows how likely an event is to occur. But what if we want a more specific answer? In many cases probability can be expressed as a fraction or a percentage. To find the probability, we compare the number of favorable outcomes (the result we are looking for) with the total number of possible outcomes.

Look at the example at the top of the page. In this case, the favorable outcome is for Lin and Jon to pull a purple marble out of the jar. There is only one case where that will happen. There is one favorable outcome. Now count the total number of outcomes. There are four marbles all together. So there are four possible outcomes.

The probability of an event can be expressed in words.

There is a one-in-four chance that Lin and Jon will pull out a purple marble.

The probability of an event can be expressed as a fraction in simplest form.

Favorable outcomes = 1

Possible outcomes = 4

There is a $\frac{1}{4}$ chance that Lin and Jon will pull out a purple marble.

The probability of an event can be expressed as a percentage.

$1 \div 4 = .25 = 25\%$

There is a 25% chance that Lin and Jon will pull out a purple marble.

© Shell Education   #50755—Leveled Texts for Mathematics: Data Analysis and Probability

Finding the Probability

Use this spinner for the examples below. Find the probability that the spinner will land on blue.

Step 1: Find the possible number of favorable outcomes for the condition of your problem.

Count the blue spaces. We find that there are 5.

Step 2: Find the number of possible outcomes.

Count the total spaces. We find that there are 10.

Step 3: Find the fractional probability. Express it in simplest form.

$$\frac{5}{10} = \frac{1}{2}$$

There is a probability of $\frac{1}{2}$ that the spinner will land on blue.

Step 4: Find the percentage probability.

$$5 \div 10 = 0.5 = 50\%$$

There is a 50% probability that the spinner will land on blue.

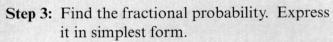

Probability Experiments in Our Daily Lives

Have you ever played a card game? How about a game with spinners or dice? Maybe you have played a game that uses special dice that have 20 sides. Some games even use dice with 100 sides! These games are all based on probability. The game designers know the exact chance for every possible outcome. They use this knowledge to try to make the game fun and fair for everyone. Video game designers use this knowledge, too. But in video games, the computer takes care of the probability. The players may never see it happen!

You Try It

Use the spinner from the example above to answer these questions.

What is the probability that the spinner will land on yellow? Express your answer as a fraction.

What is the probability that the spinner will land on pink? Express your answer as a percentage.

138

Probability Experiments

Lin and Jon are playing a game together. There are four marbles inside a jar—one marble is purple and three are yellow. If they extract a marble without looking, what is the probability that the marble they retrieve will be purple?

Basic Facts

Probability shows how likely an event is to occur. But what if we require a more specific or exact answer? In many cases probability can be expressed as a fraction or a percentage. To find the probability, we compare the number of favorable outcomes (the result we are looking for) with the total number of possible outcomes.

Look at the example at the top of the page. In this example, the favorable outcome is for Lin and Jon to pull a purple marble out of the jar, but there is only one situation where that will happen. Therefore, there is only one favorable outcome. Now count the total number of possible outcomes. There are four marbles all together, so there are four possible outcomes.

The probability of an event can be expressed in words.

There is a one-in-four chance that Lin and Jon will pull out a purple marble.

The probability of an event can be expressed as a fraction in simplest form.

Favorable outcomes = 1   Possible outcomes = 4

There is a $\frac{1}{4}$ chance that Lin and Jon will pull out a purple marble.

The probability of an event can be expressed as a percentage.

$1 \div 4 = .25 = 25\%$

There is a 25% chance that Lin and Jon will pull out a purple marble.

139

Finding the Probability

Use this spinner to help you visualize these examples. Find the probability that the spinner will land on blue.

Step 1: Determine the possible number of favorable outcomes specific to your problem's conditions.

Count the blue spaces. We determine that there are 5.

Step 2: Determine the number of possible outcomes.

Count the total spaces. We determine that there are 10.

Step 3: Determine the fractional probability and express it in simplest form.

$$\frac{5}{10} = \frac{1}{2}$$

There is a probability of $\frac{1}{2}$ that the spinner will land on blue.

Step 4: Find the percentage probability.

$$5 \div 10 = 0.5 = 50\%$$

There is a 50% probability that the spinner will land on blue.

Probability Experiments in Our Daily Lives

Have you ever played a card game or a game with spinners or dice? Maybe you have played a game that uses special dice that have 20 sides. Some games even use dice with 100 sides! These games are all based on probability. The game designers know the exact chance for every possible outcome and so they use this knowledge to try to make the game fun and fair for everyone. Video game designers use this knowledge too, but in video games, the computer takes care of the probability. The players may never see it happen!

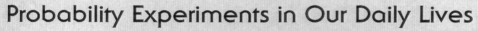

You Try It

Use the spinner from the example above to answer these questions.

What is the probability that the spinner will land on yellow? Express your answer as a fraction.

What is the probability that the spinner will land on pink? Express your answer as a percentage.

140

References Cited

August, D. and T. Shanahan (Eds). 2006. Developing literacy in second-language learners: Report of the National Literacy Panel on language-minority children and youth. Mahwah, NJ: Lawrence Erlbaum Associates, Inc.

Common Core State Standards Initiative. 2010. *The standards: Language arts.* (Accessed October 2010.) http://www.corestandards.org/the-standards/languagearts.

Marzano, R., D. Pickering, and J. Pollock. 2001. *Classroom instruction that works.* Alexandria, VA: Association for Supervision and Curriculum Development.

Tomlinson, C.A. 2000. *Leadership for Differentiating Schools and Classrooms.* Alexandria, VA: Association for Supervision and Curriculum Development.

Vygotsky, L.S. 1978. *Mind and society: The development of higher mental processes.* Cambridge, MA: Harvard University Press.

Contents of Teacher Resource CD

NCTM Mathematics Standards

The National Council of Teachers of Mathematics (NCTM) standards are listed in the chart on page 20, as well as on the Teacher Resource CD: *nctm.pdf*. TESOL standards are also included: *TESOL.pdf*

Text Files

The text files include the text for all four levels of each reading passage. For example, the Collecting Data text (pages 21–28) is the *collecting_data.doc* file.

PDF Files

The full-color pdfs provided are each eight pages long and contain all four levels of a reading passage. For example, the Collecting Data PDF (pages 21–28) is the *collecting_data.pdf* file.

Text Title	Text File	PDF
Collecting Data	collecting_data.doc	collecting_data.pdf
Creating Pictographs	creating_pictographs.doc	creating_pictographs.pdf
Analyzing Pictographs	analyzing_pictographs.doc	analyzing_pictographs.pdf
Creating Bar Graphs	creating_bar.doc	creating_bar.pdf
Analyzing Bar Graphs	analyzing_bar.doc	analyzing_bar.pdf
Creating Line Graphs	creating_line.doc	creating_line.pdf
Analyzing Line Graphs	analyzing_line.doc	analyzing_line.pdf
Creating Circle Graphs	creating_circle.doc	creating_circle.pdf
Analyzing Circle Graphs	analyzing_circle.doc	analyzing_circle.pdf
Comparing Graphs	comparing.doc	comparing.pdf
What Does *Mean* Mean?	mean.doc	mean.pdf
Median in the Middle	median.doc	median.pdf
Mode and Range	mode_range.doc	mode_range.pdf
Probability of Events	probability_events.doc	probability_events.pdf
Probability Experiments	experiments.doc	experiments.pdf

JPEG Files

Key mathematical images found in the book are also provided on the Teacher Resource CD.

Word Documents of Texts

- Change leveling further for individual students.
- Separate text and images for students who need additional help decoding the text.
- Resize the text for visually impaired students.

Full-Color PDFs of Texts

- Create overhead transparencies or color copies to display on a document projector.
- Project texts on an interactive whiteboard or other screen for whole-class review.
- Read texts online.
- Email texts to parents or students at home.

Leveled Texts
for Mathematics
Data Analysis and Probability

For use
with either
Macintosh®
or Windows®

SHELL
EDUCATION
SEP XXXXX

Teacher Resource CD

JPEGs of Mathematical Images

- Display as visual support for use with whole class or small-group instruction.

Notes

Notes